Land Tenure, Boundary Surveys, and Cadastral Systems

Land Tenure, Boundary Surveys, and Cadastral Systems

George M. Cole and Donald A. Wilson

CRC Press
Taylor & Francis Group
Boca Raton London New York

CRC Press is an imprint of the
Taylor & Francis Group, an **informa** business

CRC Press
Taylor & Francis Group
6000 Broken Sound Parkway NW, Suite 300
Boca Raton, FL 33487-2742

Printed on acid-free paper
Version Date: 20160321

International Standard Book Number-13: 978-1-4987-3165-2 (Hardback)

Library of Congress Cataloging-in-Publication Data

Names: Cole, George M., author. | Wilson, Donald A., 1941- author.
Title: Land tenure, boundary surveys, and cadastral systems / George M. Cole and Donald A. Wilson.
Description: Boca Raton : Taylor & Francis, 2016. | "A CRC title." | Includes bibliographical references and index.
Identifiers: LCCN 2016012700 | ISBN 9781498731652 (alk. paper)
Subjects: LCSH: Land tenure--Law and legislation--United States. | Land titles--Registration and transfer--United States. | Boundaries (Estates)--United States. | Surveying--Law and legislation--United States.
Classification: LCC KF574 .C65 2016 | DDC 346.7304/32--dc23
LC record available at http://lccn.loc.gov/2016012700

Visit the Taylor & Francis Web site at
http://www.taylorandfrancis.com

and the CRC Press Web site at
http://www.crcpress.com

Printed and bound in the United States of America by Publishers Graphics, LLC on sustainably sourced paper.

para los estudiantes del Colegio, ayer, hoy y mañana

Contents

Authors

George M. Cole is a professional surveyor and engineer with over 50 years of experience as a land surveyor. Beginning his career conducting hydrographic surveys of the Chukchi Sea as a commissioned officer of the U.S. Coast & Geodetic Survey, he has had a wide variety of assignments, including participation in the nation's first satellite geodesy program, as the State Cadastral Surveyor for Florida, and as president of a surveying and mapping firm with projects in both the United States and Latin America. Recently, he has concentrated on surveying education with service as a visiting professor at the University of Puerto Rico (which was the inspiration for this book) and currently as an adjunct professor at both the University of Puerto Rico and Florida State University. He has made significant contributions to professional literature and is the author of several textbooks and numerous technical papers. He also has frequently testified on surveying matters before local, state, and federal courts, including before a special master of the U.S. Supreme Court. Cole holds a bachelor of science degree from Tulane University as well as master of science and doctor of philosophy degrees from Florida State University.

Donald A. Wilson, president of Land & Boundary Consultants, has been in practice for over 50 years, consulting to groups throughout the United States and Canada. He is both a licensed land surveyor and professional forester, having conducted more than 500 programs on a variety of topics, including description interpretation, boundary evidence, law, title problems, and forensic procedures. Mr. Wilson has more than 200 technical publications in several areas and has been involved with over 50 books, which include titles on Maine history and several books on fishing. Besides being coauthor of *Evidence and Procedures for Boundary Location* and *Boundary Control and Legal Principles,* he is author of *Deed Descriptions I Have Known…But Could Have Done Without, Easements and Reversions, Interpreting Land Records,* and *Forensic Procedures for Boundary and Title Investigation.* His latest title is *Easements Relating to Land Surveying and Title Examination.* Mr. Wilson is an instructor for VectorLearning's online professional courses and a regular presenter in the University of New Hampshire's Professional Development Program. He is part owner of and lead instructor in Surveyors Educational Seminars and a regular seminar coordinator for the University of New Hampshire. In his professional practice, Don has testified numerous times, in a variety of courts, on boundary and title matters.

1

Introduction

> Since the dawn of history, land has been the primary source from which Man has taken the things that he needs for existence and for progress, for life in the lower margin and for luxury. Land still brings forth the grain, fruit and meat that feed him; the flax, cotton and wool that clothe him; the wood, stone and iron that house him. It supplies the fuels that warm him—the fuels that, transformed by his arts into energy, serve him in a thousand ways.
>
> **Charles Abrams (1979)**

Mankind has always had a special relationship with the land, and with good reason, since it is a primary key to our existence. As suggested in the previous quote, land serves as the basic platform for life and source of nourishment, shelter, and energy for mankind. It also serves as the basis of income for humans. Reflecting that essential nurturing role, land has often been called "Mother Earth."

As a result of the special relationship with and dependence on land, humans have long had a tendency to claim exclusive rights over tracts of land that they occupy, either jointly with other members of a tribe or community or as individuals. That tendency to claim a parcel of land as one's own and to defend it from others appears to be a natural tendency since it may also be observed with many other animal species. As an example, most people have observed a dog marking its territory. As another example, persons living in some parts of the United States have possibly experienced being threatened by mockingbirds defending the territory around their nest. Such actions are really not that far-removed from that of marking the corners of a land parcel with concrete monuments or the filing of a deed to the land at the county land records office. Both involve methods of defending a claim to land.

Early in recorded history, humans refined the process of defining and marking the limits of claimed lands by use of boundary surveys and landmarks. Based on such surveys, methods have been developed to uniquely describe parcels of land, which provides the basis for land being used as a commodity that can be bought, sold, traded, or used as equity. In addition, humans have refined the process of defending land claims by use of a legal system in lieu of a physical defense. This was eloquently recognized by philosopher John Locke (1698/1952) as follows:

> I cannot count upon the enjoyment of that which I regard as mine, except through the promise of the law which guarantees it to me. It is law alone

> which permits me to forget my natural weakness. Tis only through protection of law that I am able to enclose a field, and to give myself up to its cultivation with the sure though distant hope of harvest.

In most advanced societies today, a key element of that legal defense is the recordation or registration of land claims in a central database, or cadastral system. That recordation process provides public notice of a claim to a specific parcel of land and thus allows utilization of the legal system for defense of the claim.

Together, the processes of how land parcels are held; how they are defined, measured, and described to allow economic transactions; how they are marked to allow their use and defense; and how they are legally protected have allowed for the orderly possession and use of land. In doing so, these processes have also provided the basis for the advanced economy of most developed nations. For those living in societies with such property systems, those processes are often taken for granted and have become an almost invisible part of the landscape. Nevertheless, because of the importance of land to all aspects of human life, as well as the key role that it has in the economic system, the processes are essential to the orderly operation and economic well-being of society and therefore deserve careful study. That is the objective of this writing.

Very often, these processes—land tenure, boundary surveying, and cadastral systems—are considered separately. Nevertheless, they are very much interrelated, and none of these processes may be completely understood without an understanding of the others. This is especially true considering the advent of georeferenced cadastral maps reflecting the location of land parcels relative to many other components of the physical and legal infrastructure. Therefore, the interrelationship of these processes is a central concept that will be found throughout this writing.

The objective of this book is to provide an introduction to land tenure, cadastral systems, and boundary surveying, including an understanding of the interrelationship of these areas and their role in land tenure and real property law. Although intended as a basic text for college-level surveying courses, this book should also be of significant value to cadastral mappers, real property attorneys, land title professionals, and others involved with land transactions.

When used as a text for a course in the subject matter, it is suggested that the course instruction include several practical exercises to supplement the text material. For example, students should perform a title search for a parcel of land. In addition, information should be obtained using a fiscal cadastre. Other exercises should include obtaining and using Public Land Survey plats and field notes from a state repository. Practical exercises in writing and interpretation of legal descriptions and the actual participation in boundary surveys should also be included for a complete learning experience. Recommended additional reading in that chapter's topic is provided at the end of each chapter.

2

Land Tenure

2.1 Territoriality

Territoriality is usually considered to be the attachment to and defense of a specific geographic area. Although often considered a characteristic of mankind, the identification of an area for exclusive use is a well-documented practice among many species of animal life. The examples provided in the Introduction involving the "marking" of territorial boundaries by dogs and the defense of nesting areas by mockingbirds illustrate two of many examples of territoriality by species other than humans. In addition to birds and dogs, territoriality has been observed in many other species. A number of mammals claim exclusive territories and mark the boundaries of those areas with self-produced scent in the form of urine or musk on stumps, tree limbs, or rocks. Among some social animals such as the primates, there is a tendency for territory to be claimed by groups rather than individuals. Organized bands of monkeys and apes often occupy and defend territories as a group.

Therefore, territoriality is apparently a natural tendency since many examples of territoriality may be observed among various species of animal life. In many cases, this tendency takes the form of a territory being claimed and defended by an individual or close family unit. In other cases, territory is claimed and defended in common by an expanded tribal group.

Among humans, groups of people have been seizing land from other groups as far back as historical records exist. Even early cave-dwellings societies have been reported to have individual partitions within their caves. Among the early hunting/gathering societies, land was probably not considered a commodity as in most contemporary societies. Nevertheless, a particular area of land was often associated with a tribe or group of people who lived upon it and defended it. Among Native Americans, for example, a tribe usually claimed a wide area common to all members, which was defended against intruders. In addition, individual occupancy of garden patches within that territory was respected in some areas. Some groups, such as the Sitkans of Alaska, divided productive areas of the coastline among families and used cornerstones and stakes to demarcate the boundaries of those individual tracts. As other examples, the Montagnais tribe of Canada is reported to have produced good maps of their territory by carving on pieces of birch bark. Where urban societies or formal agriculture developed, both requiring intensive land use, recognition of individual rights over territory generally developed as opposed to tribal or group rights.

Thus, territoriality appears to be a natural pattern among many species of animal life, including humans. Furthermore, exclusive claim to land appears to be a universal trait of human culture and a part of the basic fabric of human society.

2.2 Land Tenure and Its Development

In human society, territoriality is manifested by the process of *land tenure*. That term may be very simply defined as the process of holding land.* Slightly more detailed, it may be said to be the relationship, whether defined legally or customarily, among people, as individuals or groups, with respect to land. Land tenure is an institution, that is to say, it is rules invented by societies to regulate behavior. The rules of tenure define how property rights to land are to be allocated with societies. It is an essential part of most legal systems and is an important part of social, political, and economic structure. Because of that multidimensional nature, it has important social, technical, economic, legal, and political aspects.

Reportedly, the earliest urban society, as well as earliest formal agriculture, developed about 8000 BC. Our knowledge of land tenure during that period is limited, but it is presumed that some sort of individual or group exclusive rights to land were recognized due to the nature of land use by those societies.

In the early stages of civilization, most forms of land tenure were collective. When people began to cultivate or otherwise intensely use the land, individual rights began to be recognized. With intensive use, it became more important to delineate and defend the areas involved to protect one's investment in the land. Eventually, this led to the development of ways to describe the land and legal systems to protect interests in land other than with physical defense, as well as to allow for the use of land as a commodity.

Definite evidence of private ownership of land and sales of privately owned land parcels in various societies has been found for periods as early as 2700 BC (Powelson 1988). That evidence is in the form of textual information on tablets in Southern Mesopotamia. No evidence exists of an organized

* The origin of the word *tenure* is the Latin verb *tenere*, which means to hold or comprehend. It is also seen in the Spanish verb *tener* or French *tenir*, meaning "to have."

system of land distribution or of concepts such as title deeds, but land was apparently claimed and occupied by those who would take it and hold it.

Egypt provides another example of an early society developing sophisticated land tenure practices. The oldest documented evidence of such a system is dated approximately 2350 BC. The land system in that country, even then, had laws requiring a document to be drawn up on papyrus, signed by three witnesses, and stamped with an official seal when transferring ownership of land. Furthermore, that system accounted for lands by use of a central registry in the office of the vizier, the pharaoh's prime minister. That registry also was used for recordation of wills and other documents affecting land ownership similar to the process in use in advanced societies today.

An interesting example of the relative complexity of early land transactions is one conducted by the prophet Jeremiah, which may be found in *Holy Bible* as follows (Jeremiah 32: 9–14, *Holy Bible*, Revised Standard Edition)

> And I bought the field at An'athoth from Han'amel my cousin, and weighed out the money to him, seventeen shekels of silver. I signed the deed, sealed it, got witnesses, and weighed the money on scales. Then I took the sealed deed of purchase, containing the terms and conditions, and the open copy; and I gave the deed of purchase to Baruch the son of Neri'ah son of Mahsei'ah in the presence of Han'amel my cousin, in the presence of the witnesses who signed the deed of purchase, and in the presence of all the Jews who were sitting in the court of the guard.
>
> I charged Baruch in their presence, saying, Thus says the Lord of Hosts, the God of Israel: Take these deeds, both this sealed deed of purchase and this open deed, and put them in an earthenware vessel, that they may last for a long time.

As may be seen, the transaction involved many of the same elements used today, including the requirement for witnesses to the transaction, public notice of the sale, and the preservation of the deed in a safe place such as a sealed vessel as was the custom of those days.

During the dynasty of Ptolemy (305–31 BC), a relatively advanced land information system was developed in Egypt that contained precise information on even small plots of land, including the legal status of the land, its dimensions, its location, the holder of the land, the state of its irrigation, and the type of crops grown on it.

It may be seen from the foregoing examples that having exclusive rights to land is a long-established practice of mankind. Furthermore, it may be seen that many of the practices associated with land ownership today, such as considering rights to land as a commodity that can be traded, the creation of formal land title deeds to represent land in transfers of ownership, the maintenance of a central registry of deeds to land, the recordation of other documents affecting rights to land such as wills, and even land information systems, are also long-established and essential components of human society.

As various societies have evolved, various degrees of private rights to land were recognized. At times, in some regions, land ownership became consolidated under a sovereign power. At other times in history, individual rights to land were recognized. Nevertheless, the general tendency in most urban or agrarian societies, especially in more recent times, has been to move toward individual land ownership rights with the demise of feudalism.

2.3 Modes of Interests and Tenure in Land

Practices and laws regarding land tenure vary considerably among nations. Nevertheless, there are some common principles. Generally, most systems provide for several levels of interest in land, some of which may intersect one another and exist simultaneously. For example, under common law, *fee simple* is the most complete ownership interest one can have in real property, other than the rare *allodial title*, when real property is owned absolutely free and clear of any superior landlord or sovereign. Yet, even land held in fee simple may be subject to overlapping interests, such as when the property is leased or rented by its owner to another party. Generally, when there is more than one interest in property, the interests are considered to be in one of the following classifications:

Overriding interests: when a sovereign power (e.g., a nation or community) has the powers to allocate or reallocate land through expropriation or similar practice

Overlapping interests: when several parties are allocated different rights to the same parcel of land (e.g., one party may have lease rights, and another may have a right of way). Examples of overlapping interests are in abundance. An obvious one is the land, burdened by a highway easement, further burdened by slope and sight easements, which in turn are burdened by overhead and underground utilities, along with telephone and cable lines. In major metropolitan areas, easements are stacked on top of one another, some being surface, while others are subsurface and above surface (air space)

Complementary interests: when different parties share the same interest in the same parcel of land (e.g., when members of a community share common rights to grazing land, community wells, and the like)

Competing interests: when different parties contest the same interests in the same parcel (e.g., when two parties independently claim rights to exclusive use of a parcel of agricultural land. Land disputes arise from competing claims)

Many different forms of land tenure can be found within a given society. *Rights in common*, especially for many agricultural uses, may be held in trust for the citizens. *Private tenure* is where rights to land are assigned to a private entity, which may be an individual, a husband and wife, a corporation, or other groupings. *Communal tenure* is when rights are assigned to a community or group where any member of that group has the right of use. *Open access tenure* is when rights are assigned to no one in particular and no one can be excluded. *Governmental tenure* is when rights are assigned to some authority in the public sector. For any of these types of land tenure, rights may include exclusive rights or more limited rights such as leasehold. Multiple rights may be held by several different persons or groups with differing rights in the same parcel of land frequently held by different parties. Thus, *tenure* is a broad and multidimensional concept. In common language, this has given rise to what is known as "the bundle of rights."

Rights in land may be simplified by categorization into use rights, control rights and transfer rights. Rights may also be classed as formal or informal. Formal property rights are those acknowledged by the state and which may be protected through legal means. Informal property rights lack official recognition and protection.

Property rights may also be illegal under some circumstances. Often, illegal property holdings arise because of inappropriate laws. In other cases, property may be classed as extralegal. These are situations that are not against the law but are also not recognized by the law. It is important to distinguish between statutory rights, or "formally recognized rights," and customary rights, or "traditional rights." Custom, or customary rights, may play an important role in identifying rights of a group of people established over a long period of time. Customary rights in land involving indigenous societies are usually created following their traditions and through the ways in which community leaders assign land use rights to the members of the community. They are often rights developed through ancestral occupation and use by ancestral societies.

There are a variety of modes of land ownership and tenure. Some involve individual ownership and others involve collective ownership, which may take the form of membership in a cooperative or shares in a corporation, which owns the land (typically in fee simple, but possibly under other arrangements). There are also various hybrids, such as in some communist states, where government ownership of most agricultural land is combined in various ways with tenure for farming collectives.

Allodial title—Allodial title is a system in which real property is owned absolutely free and clear of any superior landlord or sovereign. True allodial title is rare, with most property ownership in the common law world (Australia, Canada, Ireland, New Zealand, United Kingdom, and United States) being in fee simple. Allodial title may be conveyed, devised, gifted, or mortgaged by the owner, but it may not be distressed and restrained for collection of taxes or private debts or condemned by the government through eminent domain proceedings.

Common areas in colonial states—In the Pennsylvania case of *Bruker v. Burgess and Town Council of Borough of Carlisle,** the court found some of the evidence confusing, and incomplete in its wording, but stated that ancient rights as established in favor of the public continue; however, in the absence of more complete definition, uses may change over time to meet the needs of the community.

Two things in this case are noteworthy. The first is the court's statement that "there is no doubt that the mere fact of the use of (the Square) by the public for now more than 200 years is sufficient to raise a conclusive presumption of an original grant for the purpose of a public square; such is an ancient and well established principle of the law. Nor can it be denied that, where such a dedication has been established and the public has accepted it, there cannot be any diversion of such use from a public to a private purpose, and it is also true that, where a dedication is for a limited or restricted use, any diversion therefrom to some purpose other than the one designated is likewise forbidden" (citing several cases).

The second item of note is the dissenting opinion (of two of the judges) which states, in part

> Evidence of dedications and titles based upon ancient documents or events arising out of antiquity cannot possibly be as strong or clear as would be required in matters arising or titles created in modern times; and we must not lightly strike down rights or public uses which have existed for more than a century. Counsel for the Borough frankly admits that if the Borough can change or destroy this market place it can also change or destroy the church which was built on this same public square and like the market place has been used for more than 200 years. This church and this market place were dedicated by Thomas Penn in 1751; and this dedication and use as a market place were thereafter frequently ratified. The Borough merely contends that there exists today no clear evidence that this square was the 'spot' which was dedicated for these purposes.
>
> We are living in exciting and rapidly changing times. We rode from the horse and buggy age to the automobile age and then flew too rapidly to the airplane age and the atomic age. The tremendous changes which have occurred and are still daily occurring have necessarily produced uncertainty, unrest and confusion—not only in the minds of men, but in many phases of man's life. As a consequence, 'change' is on every man's lips, and unrest and uncertainty in many a man's heart. In the craving for change, in the restless quest for a Utopia of riches and ease, haven't we too often forgotten the things of the spirit, as well as the history of our Country and the immemorial customs of our people? Haven't we rushed frantically and heedlessly after false goods—material prosperity and

* 376 Pa. 330; 102 A.2d 418, Pennsylvania (1954)

> political panaceas? Should not our wonderful farm people be allowed to preserve in a farming community a few of their ancient privileges, customs and practices, even though more income would be produced by a different (public or private) use?
>
> Title to this market place or square is not in the Borough of Carlisle; it is in the Commonwealth of Pennsylvania, with a reversionary interest in Penn's heirs. The Borough has no title or estate in this property; it has only a right to control and regulate, in the interest of the public, the public use of the 'market place' for the uses and purposes to which it was dedicated.*

An *easement* is a right of use of land owned by others for a designated purpose, but not ownership. The owner of the land may continue to use the land over which there is an easement as long as it does not interfere with the use of the land specified in the easement. Easements are typically used to provide formal permission for utility lines to cross over or under property and for providing ingress and egress to adjacent property. The most classic easement is the right of way.

Fee simple estate—This is exclusive ownership interest in property. The owner(s) may hold, convey it, and pass it on to heirs without limitation.

Fee tail—Under common law, this is hereditary, nontransferable ownership of real property. A similar concept, the *legitime,* exists in civil and Roman law. The legitime limits the extent to which one may disinherit an heir.

Feudal land tenure—Feudal land tenure is a system of mutual obligations under which a royal or noble personage is granted a fiefdom (some degree of interest in the use or revenues of a given parcel of land) in exchange for a claim on services such as military service or simply maintenance of the land in which the lord continued to have an interest. This pattern obtained from the level of high nobility as vassals of a monarch down to lesser nobility whose only vassals were their serfs.

Leasehold or rental—Under both common law and civil law, land may be leased or rented by its owner to another party. A wide range of arrangements is possible, ranging from very short terms to the 99-year leases common in the United Kingdom, and allowing for various degrees of freedom in the use of the property. Generally, improvements to the land during the term of the lease, such as buildings, remain the property of the owner of the property.

A *lien* is a claim on property for payment of a debt or obligation. It is not the right of possession but is the right to have the property sold to satisfy the debt. Liens are frequently used by contractors as a means of collecting unpaid fees for work performed on improvements to property. Government entities may also impose liens because of failure to pay taxes on the property.

Life estate—Under common law, this is an interest in real property that ends at the holder's death. The holder has the use of the land for life but typically no ability to transfer that interest or to use it to secure a mortgage loan.

* *Hoffman v. City of Pittsburgh,* 365 Pa. 386, 75 A.2d 649 (1950)

In those cases where such estate does pass to another, the next holder is said to have an estate *per autre vie* (for the life of another). It, too, terminates upon the death of the original holder.

A *mortgage* is a conditional conveyance of real property as a pledge for the security of a debt. An example of a mortgage is when a person borrows money to purchase property. That person guarantees that he or she will repay the money by making a conditional conveyance to the lender. If the money is repaid as specified, the mortgage becomes null and void. If the money is not repaid as specified, the lender may take possession of the property.

Native title (aboriginal title), also referred to as indigenous title, original Indian title, and customary title, is a common law concept that recognizes that some indigenous people have certain land rights that derive from their traditional laws and customs. Native title can coexist with nonindigenous proprietary rights, and in some cases, different indigenous groups can exercise their native title over the same land. In the United States, aboriginal title is a common law doctrine that the land rights of indigenous peoples to customary tenure persist after the assumption of sovereignty under settler colonialism. It has been described as an aboriginal right, or the right to the land itself, and is a communal right. Nearly all jurisdictions are in agreement that aboriginal title is inalienable, except to the national government, and that it may be held either individually or collectively. (See Section 3.1.7 for expanded information on this topic.)

Public trust easement—The ancient laws of the Romans held that the seashore not appropriated for private use was open to all. This principle became the law in England, and these rights were further strengthened by later laws in England and subsequently became part of the common law of the United States.

An easement through public trust was one of the claims made in the case of *Akau v. Olohana Corp.** In this case, the plaintiffs had lived or fished in Kawaihae for many years. They represented two subclasses: one containing Hawaii residents who used or were deterred from using the trails, and the other containing all persons who own land or reside in the area and used or were deterred from using the trails. The defendants were landowners or tenants who possessed the beachfront land between Spencer Beach Park and Hapuna Beach Park, a span of about two and a half miles along the beach. They had barred all public access across their land to the public beach since acquiring the land in 1954.

Two of the trails in issue run roughly parallel to the beach between the two parks and had existed since before the turn of this century. The Kamehameha Trail is at most points very close to the water and at others about 100 yards away. The Kawaihae-Puako Road is about 150 yards further upland. There are also 11 intersecting trails that run from the main trails to the shore. Plaintiffs allege that these trails had been used by the public until 1954.

* 65 Haw.383, 652 P.2d 1130, Hawai'I (1982)

The Territory of Hawaii owned the land between the two parks until it was sold to Richard Smart in 1954. The parcel consisting of the Kawaihae-Puako Road was also sold to Smart at that time. Smart conveyed all his land by deed or lease, and all the original defendants were owners or lessees of that land.

Plaintiffs claimed that the trails have been and are public rights-of-way and asked for declaratory and injunctive relief to that effect. The eight theories plaintiffs rely on are (1) HRS §7-1; (2) ancient Hawaiian custom, tradition, practice, and usage; (3) common law custom; (4) easement by implied dedication; (5) easement by prescription; (6) easement by necessity; (7) easement by implied reservation; and (8) easement through public trust. The State was made a nominal defendant to protect the interests of the public. Its position, however, is in support of the plaintiffs in the appeal.

It is noted that the *Akau v. Olohana Corp.* case had to do with class standing as opposed to the State bringing the action. Yet, it is illustrative of the theories under which an action of this nature may be brought.

In the case of *Banner Milling Co. v. State,** the New York court explored the differences in land rights held by a state. It said, "We are of the opinion that the vital question in this case is this: Was the land, the title of which is in question here, owned and held by the State as a sovereign in trust for the People, or as a proprietor only? There is a well-recognized distinction between lands held by the State as sovereign in trust for the public and lands held as proprietor only, for the purpose of 'sale or other disposition.' (*Weber v. State Harbor Comrs.*, supra, 18 Wall. 68) In either circumstance, except a statute (making an agreement on behalf of the People not to sue) authorizes it, lands of a sovereign State cannot be lost to, or taken from, the State by failure to assert her title (2 C.J. 213; *Fulton L., H. & P. Co. v. State,* 200 N.Y. 400; *St. Vincent F.O. Asylum v. City of Troy,* 76 id. 108; Hays v. U.S., 175 U.S. 248); and, after such a statute has been passed by a State, such lands only as the State holds as a proprietor may be lost to the State; it cannot lose such lands as it holds for the public, in trust for a public purpose, as highways, public streams, canals, public fair grounds. (*Burbank v. Fay,* 65 N.Y. 57; 2 C. J. 213, 214, 215)."

Rights to use a common—These may include such rights as the use of a road or the right to graze one's animals on commonly owned land.

Sharecropping—Under sharecropping, one has use of agricultural land owned by another person in exchange for a share of the resulting crop or livestock.

Traditional land tenure—Most indigenous nations or tribes of North America had no concept or formal notion of land ownership. When Europeans first came to North America, they sometimes disregarded traditional land tenure and simply seized land, or they accommodated traditional land tenure by recognizing it as aboriginal title. This theory formed the basis for treaties with indigenous peoples.

---

* 117 Misc. 33, New York (1921)

2.4 Conveyance of Interests in Land

There are a number of processes by which legal interests in land may be conveyed by the owner to others. Some of the most utilized processes include deeds, dedications, and wills.

Deeds are the most common and the most important instrument used for conveyance of interest in real property. A deed is always a formal document and, by law, must be written. There are two principal types of deeds: warranty deeds and quitclaim deeds. Warranty deeds are instruments whereby the grantor warrants that he or she is the lawful owner of the property being conveyed and is therefore bound to forever defend the title to the property. With quitclaim deeds, the grantor does not warranty that he or she owns the land being conveyed. Rather, a quitclaim deed is used to convey whatever interest that the grantor has in the property to the grantee. Another very specific type of deed is a patent. It is a deed from a sovereign (or sovereign government) to an individual. Examples of this are the many patents used to convey parcels of the public lands of the United States to individuals.

A dedication is a conveyance of real property or rights to real property to the public. An example of this is the dedication of the streets in a subdivision to the public as part of the formal recordation process. The interests conveyed by such a dedication must be accepted by the government entity named as the grantee for the dedication to be valid.

An additional means of conveying interest in land is by means of a will. A will is a declaration of a person's wishes for the distribution of his or her property after death. Before property ownership can be actually transferred under the terms of the will, it must be submitted to a probate court for approval.

2.5 Transfer of Interests in Land without Agreement of Owner

In addition to the various means by which interest in land may be conveyed by the owner to other individuals or the public, there are several means by which interest in land is conveyed *without* the agreement of the owner. The most utilized of these processes include condemnation under the right of eminent domain, adverse possession, and prescription.

The public may acquire an interest in land from an individual through *condemnation* procedures under the right of eminent domain. When this process is utilized, the transfer is involuntary on the part of the private land owner. An example of this may occur when a state or county is constructing a project in the public interest, such as a new roadway, through lands where the owner is not willing to sell at the offered price. Condemnation requires a judicial finding to approve the process and to establish fair compensation to the owner for the property as well as for consequential damages.

Adverse possession is the acquisition of title to real property without the agreement of the owner based on long-continued possession of the property by another. The right to acquire property by this means is also called *squatter's rights.* Requirements for acquiring property in this manner vary with state law but typically require open and exclusive possession of the property for a certain period of time in such a manner that the owner would be aware of the possession. The ultimate effect of adverse possession is to extinguish the former title. The following two definitions best describe the significance of the adverse possession process:

> Adverse possession is the ripening of hostile possession, under proper circumstances, into title by lapse of time. (3 Am Jur 2d Adverse Possession, §2)
> Adverse possession generally creates an absolute title to real property in fee simple, which is as good as title by patent from the state or title by deed from the record owner, although it does not amount to record title unless and until made so by judicial proceeding. (3 Am Jur 2d Adverse Possession, §1)

In the United States, there is a long history of property being acquired by adverse possession. In colonial times and even after the American Revolution, occupying land without title, or "squatting," was a frequent practice in the United States. Even George Washington, the first president of the United States, is known to have had problems with squatters on his Virginia farmland and was once warned by his lawyer that "if he succeeded in his suit against squatters on his estate, they would probably burn his barns and fields" (de Soto 2000). Gradually, this situation was improved as public domain land was made more readily available to settlers and land recordation systems became more efficient and accessible.

Prescription is a method of acquiring easement rights to property without the agreement of the owner through long-continued usage. An example of this could occur if you acquired property where the previous owner had been allowing his neighbors to cross his property to access their property for many years. The neighbor could possibly have acquired prescriptive rights to continue use of the property for a driveway that would preclude you from closing the property to his use after buying the property. Another application of this process allows the public ownership of road right of way over lands used and maintained by the public for a long period of time.

Custom is a method of acquiring easement rights by the operation of law through continuous use of land for a specific purpose over a long period of time. An easement right acquired by custom lacks a dominant tenement and is in generally in favor of the inhabitants of a particular locality.

Custom differs from prescription, which is personal and is annexed to the person of the owner of a particular estate, while the other is local and relates to a particular district. The distinction has been thus expressed: *"While prescription is the making of a right, custom is the making of a law."** Bouvier's Law Dictionary defines custom as "such a usage as by common consent and uniform practice has become the law of the place, or of the subject matter to which it relates." Another description of the distinction is provided by an early New Hampshire case as follows:

> Whether rights are holden as a custom, or as a prescription, depends on the manner in which they are holden—whether by a local usage, or as a personal claim, or as dependent on a particular estate. All rights which may be holden as a custom, may be holden as a prescription, but the reverse is not true. If the rights are common to any manor, district, hundred, parish or county, as a local right, they are holden as a custom. If the same rights are limited to an individual and his descendants, to a body politic and its successors, or are attached to a particular estate, and are only exercised by those who have the ownership of such estate, they are holden as a prescription.†

2.6 Land Tenure in Contemporary Society

As previously mentioned, the general trend in land tenure in recent years has been toward individual land ownership. That type of land tenure, protected by a formal land property system, has been a key factor in the prosperous economic system of the United States and other economically advanced nations. This is due to the ability of owned land to have an economic life in addition to its physical life. Such a system provides security of title to a parcel of land, thus encouraging stable development. In addition, the land becomes an asset that can be traded to take advantage of increased value and one that can be used as security for mortgages. The land also represents an accountable address for collection of taxes and debts and provides a basis for the development of reliable public utilities.

Yet, in many developing nations in the so-called third world, large percentages of people reside on land without formal title. In some areas, massive squatter communities, called *favelas* in Brazil, *ranchos* in Venezuela, *barrios marginales* in Mexico, *pueblos jóvenes* in Peru, and *paracaidistas* in Central America, exist. De Soto (2000) estimated that 80% of all real estate in Latin America is held extralegally. Although such informal communities represent a tremendous amount of financial investment, the lack of formal property title prevents the occupied land from contributing to the economy. This has

* Lawson, Usages & Cust. 15, note 2

† *Perley v. Langley*, 7 N.H. 233, New Hampshire (1834)

led some researchers to the conclusion that such extralegal land occupation is a major obstruction to prosperity for the nations in which it exists.

During the 20th century, there was a movement in some regions of the world away from individual land ownership associated with the communism movement. That movement advocated common, or state, ownership of land. After more than a century of competition between capitalism, which supported individual ownership of land, and communism, which supported state ownership, almost all of the former communist nations have moved toward a more capitalistic economy featuring individual land ownership. This is in an apparent attempt to take advantage of the economic advantages of individual land ownership.

Thus, while the general trend of land tenure worldwide has been toward individual land ownership, there have been opposing movements at times, such as the communism movement in the 20th century, which attempted to move away from that principle. Furthermore, there are still large numbers, possibly representing a majority of the earth's population, living in communities without legal claim to the land that they occupy. Nevertheless, the general trend, supported by the apparent economic advantages of land ownership, appears to be toward more individual land ownership.

It is interesting to reflect on the concept of individual land ownership as practiced today. As previously mentioned, it does appear to be a natural tendency of mankind. Yet, on examination of history, struggles over the "ownership" of land have often been brutal. The following paragraph from a prominent land surveying textbook (Brown et al. 1981) provides much food for thought:

> Real property is not a creation of mankind. Original ownership, therefore can only come from conquest or discovery. Private ownership of land has many injustices, but perhaps it is the least of several evils. An alternative (although completely undesirable) plan would be the nationalization of land as prevails in some noncapitalistic countries. As civilization progressed, strong-armed men assumed ownership by force. All titles today run back to and are maintained by force.

2.7 Legal Restrictions on Land Ownership and Use

Absolute land ownership is a theoretical concept (Dekker 2003). Practically every society has placed certain restrictions on the use of land that is "owned" by individuals. Indeed, for almost as long as there has been individual land ownership, there has been some type of restriction on the completely free use of land imposed by rulers or governments. These restrictions may be in the form of a tribute or land tax paid to the ruler or government, restrictions on hunting or fishing, restrictions on how the land can be used, and many other restrictions. This was true even in the American colonies. As societies have become more complex, there has been increasing emphasis on the general welfare and the distribution of rights among all members of society as opposed to only the rights of the owners of land. In many ways, that increased emphasis developed is a means of achieving the goals of the socialistic movement regarding land ownership. Yet, regardless of one's social philosophy, there is no question that as neighborhoods or nations become more densely populated, the way one person uses his or her land affects others more and competition for resources increases. As a result, it is also without question that such increased density requires increased mediation by governmental policy and increased restriction on land use.

The existence of these overriding interests, some of which are public and others solely between individuals or groups of individuals, represents legally valid restrictions on use of land. Therefore, they represent a legally valid interest in land. Yet, records and information regarding most of those interests are not typically found in a lands records repository. Rather, they are scattered among many public agencies. Thus, presently, the determination of all interests in a specific parcel of land is currently not simply a matter of researching ownership records found in a cadastral system.

2.7.1 Restrictions of Public Nuisances

A great deal of land use regulation is based on the common law concept of *nuisance*, which holds that a property owner cannot use his or her property in a manner that injures that of a contiguous owner (Dale and McLaughlin 1999). Typical forms of regulated nuisances include the causing of bad odors or air pollution, blocking off a neighbor's light and air, making excessive noise, and other actions that endanger or impair the enjoyment of nearby property.

The development of modern land use regulation based on nuisance may be illustrated by a review of early regulations for the city of London. Although that city has had controls governing the way in which buildings are constructed and maintained since the 12th century (Dale and McLaughlin 1999), the first modern building law was enacted there in 1666. The Great Fire of that year raged through the city and destroyed 13,200 houses, leaving over 100,000 persons homeless. Much of the damage associated with the fire was attributed to the lack of regulation of building construction material, size and location, and access for emergencies. In the aftermath of the fire, regulations were enacted that allowed only stone and brick to be used for the exterior of buildings, put height restrictions on houses, and increased the width of streets. The law also required that public nuisances, such as breweries and tanneries, be relocated from the center of the city to more suitable outlying areas (Platt 2014).

In the United States, land rights have traditionally been highly valued. Nevertheless, limits have always been placed on such rights when they could interfere with the rights of others by creating a public nuisance. For example, as early as 1872, the U.S. Supreme Court clearly expressed this concept with a decision upholding an ordinance in New Orleans restricting livestock slaughter houses with the following language:

> Unwholesome trades, slaughter-houses, operations offensive to the senses, the deposit of powder, the application of steam power to propel cars, the building with combustible materials, and the burial of the dead, may...be interdicted by law, in the midst of dense masses of population, on the general and rational principle that every person ought to use his property as not to injure his neighbors; and that private interests must be made subservient to the general interests of the community. (Slaughterhouse Cases, 16 Wall 36, 1872)

With the increasing urbanization of the United States, regulation of nuisances has become more prevalent and increasingly restrictive. Such regulations commonly are used to restrict many types of activities perceived to be nuisances. In addition to those discussed previously, they also include noise, unmowed lawns, buildings in disrepair, and items such as signs and billboards that are aesthetically offensive.

2.7.2 Land Use Zoning

Another type of restriction on the use of land is land use zoning. This process involves dividing a community into a system of zones in which only certain types of land use are permitted. Common classes of land use for this purpose include residential, commercial, industrial, and agricultural. Often, such principal land use classifications are further subdivided into classes such as single-family residential, multiple-family residential, and so forth. For each class, zoning regulations specify which uses are permitted, which are prohibited, minimum lot size, allowable density of development, and other requirements.

Land use zoning is often called "Euclidian zoning," after the town of Euclid, Ohio, where a challenge of its zoning ordinance resulted in the original approval of zoning by the U.S. Supreme Court (*Village of Euclid v. Ambler Realty Co.**). That decision answered the basic question of whether a town or city can deny property owners the right to certain uses of their land under our laws. Following that decision, zoning has become a widely used practice.

As with most land use regulation, zoning is largely a function of local and state laws, not a federal function. Typically, communities appoint a citizen's planning commission to develop a comprehensive plan defining how the community is to be zoned, resulting in the development of a zoning map for the community. The resulting plans are then used to regulate orderly development of the community. Many recent comprehensive plans also include measures to encourage land to be used or developed in accordance with

---

* 272 U.S. 365 (1926)

public policy objectives such as reducing urban sprawl, preservation of open space, preservation of agriculture, encouragement of industry, and ensuring an adequate supply of affordable housing.

2.7.3 Subdivision Regulation

Most communities have subdivision regulations restricting the development of land. In the United States, over 60% of new residential construction is in the form of single-family houses, in contrast to the rest of the world, where much of new housing is in the form of high-rise apartments. Single-family houses are typically constructed on an individual parcel of land, usually in a subdivision of a larger tract of land. Thus, subdivision regulations are a key component of land use regulation in the United States.

Since subdivisions often include streets that are dedicated to the public, subdivision regulations usually have minimum requirements for the alignment, size, and paving of streets to prevent future maintenance problems as well as for storm water drainage. Subdivision regulations also usually address sanitary sewer, water, and electric utility requirements. In addition, these regulations typically include requirements for the survey and monumentation of the subdivision lot boundaries and requirements for the subdivision plat.

2.7.4 Private Deed Restrictions

Not all restrictions on land use are imposed by government. Deed restrictions imposed by land developers are often used to control some aspects of private subdivisions. Such restrictions are based on a voluntary contract between a developer and lot purchasers and therefore involve no public regulatory function. Often, deed restrictions are more restrictive than governmental requirements. They are often used to require certain sized homes, the external appearance of houses, landscaping and fencing, underground electric service, and other factors. Enforcement of private deed restrictions is not a governmental function. It requires legal action by the developer or by other affected land owners subject to the same restrictions.

2.7.5 Land Development and Use Regulation

Wetlands are natural areas where the groundwater table is normally at or close to the surface. Wetlands are widely recognized as among the most important ecosystems on the earth. Some of the currently recognized contributions to the environment by wetlands include plant and animal habitat, water purification, groundwater recharge, floodwater storage and flood peak reduction, aesthetic and recreational values, and carbon storage (Cole 2007).

Floodplains are natural overflow areas adjoining surface waters. While some floodplains are also wetlands, many are not, even though they are subject to flooding during storm events. Floodplains are often defined based on the probability of being flooded in any given year. As an example, the 100-year floodplain is an area having a 1% probability of being flooded in a given year or a 100% probability of being flooded within a 100-year period. In many fast-growing areas, extensive development has occurred in wetlands and floodplains, without regard to either the environmental sensitivity of the wetlands or the potential danger to development associated with floodplains. As a result, there have been costly losses and significant damage to the environment, both of which have led to changes in land use policy.

In many floodplain areas, the fertile alluvial soil is ideal for agriculture, and such use is reasonably compatible with occasional flooding (Platt 2014). The primary concerns regarding land use in floodplains are the loss of life and property from flooding that can result when houses are constructed in these areas and concern that alteration of the natural topography of the floodplain can cause worse flooding. Based on these concerns, land use regulation for floodplains has focused on reducing hazards to life and property. Generally, such regulations allow agriculture and similar low-impact use as long as no topographical alterations are made. Construction is usually not allowed in the floodway zones closest to the surface water but generally allowed in the outer portions of the floodplain with the provision that occupied areas and such facilities as septic tanks be elevated above maximum flood elevations.

In wetlands, regulations are more restrictive. The goal in those areas is to preserve the wetland in its natural condition or to restore it to that condition due to environmental considerations. Generally, all dredge and fill and construction activities are banned in wetlands.

In both floodplains and wetlands, there are typically both local and state regulations. In addition, there are also federal restrictions in both. Federal wetland regulations are based on Section 404 of the Clean Water Act and are jointly administered by the U.S. Army Corps of Engineers and the U.S. Environmental Protection Agency. Federal floodplain regulation is administered through the National Flood Insurance Program administered by the Federal Emergency Management Agency.

2.7.6 Compensation for Land Use Restrictions

Without question, private land owners hold property subject to the general right of the community to regulate land use for the public welfare. It is also without question that land use regulation may affect the value of private land. As a result, land use regulation involves a delicate balance between public and private interests and frequently provokes controversy. Land regulation can produce a paradox at times. One of its primary purposes is to protect and enhance property values. Yet, land regulation may reduce or even totally eliminate the value of property by limiting the options for which the property may be used. When private property is taken for public use under eminent domain laws, the private owner receives just compensation. Yet, with land use regulation, compensation is typically not paid to the owner even if the owner loses the option of using the land. This can create a fairness issue.

Private property rights are protected by the Fifth Amendment in the Bill of Rights of the U.S. Constitution as follows:

> No person shall be…deprived of life, liberty, or property, without due process of law; nor shall private property be taken for public use without just compensation.

As a result, a common question confronting land use regulation is to what extent regulations may reduce the value of private property without compensation to the owner due to the regulation being a "taking" of land. This issue has frequently been litigated with the general rule being stated by Justice Oliver Wendell Holmes in the U.S. Supreme Court case of *Pennsylvania Coal Co. v. Mahon** as follows:

> The general rule at least is that while property may be regulated to a certain extent, if regulation goes too far, it will be recognized as a taking.

As new forms of land use regulation have been propagated in conjunction with the increased urbanization of the United States, there have been frequent legal challenges to those restrictions. In general, judicial decisions on these issues have favored the public interest and allow increasing regulation. Yet, it has been recently held that when a regulation "denies all economically beneficial or productive use of land," it is a compensable taking (*Lucas v. South Carolina Coastal Council*†).

Recommended Additional Reading on Land Tenure

Clawson, Marion (1964). *Man and Land in the United States.* Lincoln: University of Nebraska Press.

Desoto, Hernando (2000). *The Mystery of Capital.* New York: Basic Books (Perseus Book Group).

* 260 U.S. 393 (1922)

† 112 S.Ct. 2886 (1992)

3

Land Tenure in the United States

3.1 Origins of U.S. Land Law

With investment in land resulting from intensive use, the defense of land claims became more important. In early human society, physical defense of claimed territory was necessary. But as society evolved, laws were developed that defined the concept of ownership and allowed the use of a legal system in lieu of a physical defense. Such laws varied with different regions of the world.

As the New World was settled by a variety of European powers, each brought its own system and concepts of land tenure, along with their methods of conveyancing, surveying, and terminology. This gave rise to the use of a variety of mathematical units along with the establishment of localized legal systems. As disputes arose, they were resolved by various courts in accordance with the rule in effect at the time. Some of these adjudications remain in effect today, and in determining and locating land boundaries, they must be considered in their own context.

Land law in the United States is a combination of both civil and common law. Civil law is generally *codified,* based on a comprehensive compilation of legal rules and statutes. Jurisdictions that follow that tradition have comprehensive legal codes specifying almost any matter that can be litigated. The judge's role in such a system is to establish the facts of the case and apply the provisions of the code. Thus, the judge's decision is less crucial in shaping civil law than are the decisions of legislators who draft the law. In contrast to the civil law, common law is generally *uncodified* and not based on a comprehensive collection of rules and statutes, although it does rely on some statutes. Rather, the common law is largely based on the *precedent* of judicial decisions previously made in similar cases. Judicial opinions are maintained as a body of law in yearbooks and reports. The precedent to be applied in each case is determined by the presiding judge. Thus, judges have a significant role in shaping the common law.

The two legal traditions vary considerably with location. The civil law tradition has its roots in the early Roman legal writings and spread to continental Europe. It was applied in the colonies of European imperial powers such as Spain and Portugal and was also adopted in the 19th and 20th centuries by countries formerly possessing distinctive legal traditions, such as Russia and Japan, in efforts at legal system reform in search of the economic and political power comparable to that of Western European nation-states. The common law tradition emerged in England during the Middle Ages and spread to the many British colonies across the continents. Figure 3.1 illustrates the legal traditions prevailing throughout the world.

Although the legal system of the United States generally follows the common law tradition brought to the North American colonies from England, traces of the civil law tradition and its importance in the hemisphere may be found within state legal traditions across the country. Combinations of English common law with Roman, Spanish, French, Russian, and Dutch Civil Law exist in various states (Figure 3.2). An example of this is Louisiana, where state law is based on civil law due to Louisiana's history as both a French and a Spanish territory. Many of the southwestern states also have traces of civil law influence in their state constitutions and codes due to their early legal status as territories of colonial Spain and Mexico. California has a state civil code organized into sections similar to traditional Roman civil law categories; yet, the law within California's code is mostly common law.

This diversity presents an interesting twist to traditional boundary retracement for the practicing surveyor. Since "a boundary, once established, must remain fixed in position through any series of mesne conveyances" (Griffin 1960), the original creating documents must be interpreted in accordance with the conditions and circumstances at the time. Proper rules and laws must be considered as well as correct interpretation of units of measurement. For example, in Spanish-settled regions, as well as smaller areas of Spanish

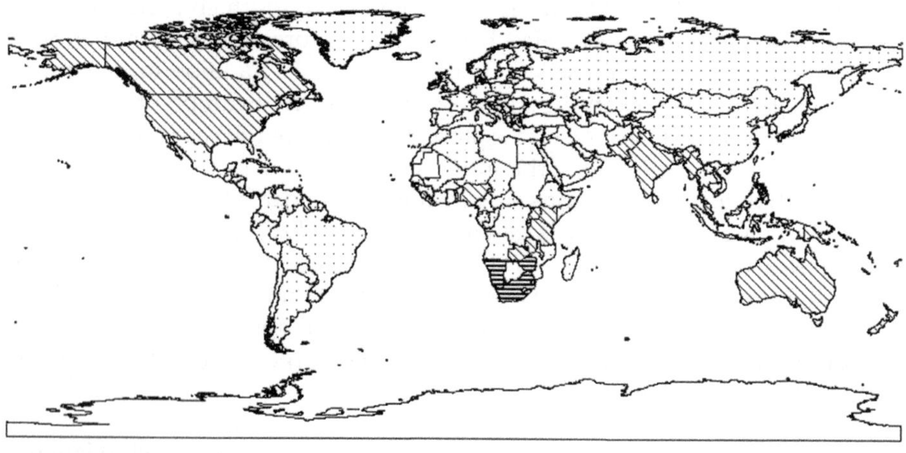

FIGURE 3.1
Worldwide legal traditions. Dots indicate common law, diagonal stripes indicate civil law, horizontal stripes indicate a combination.

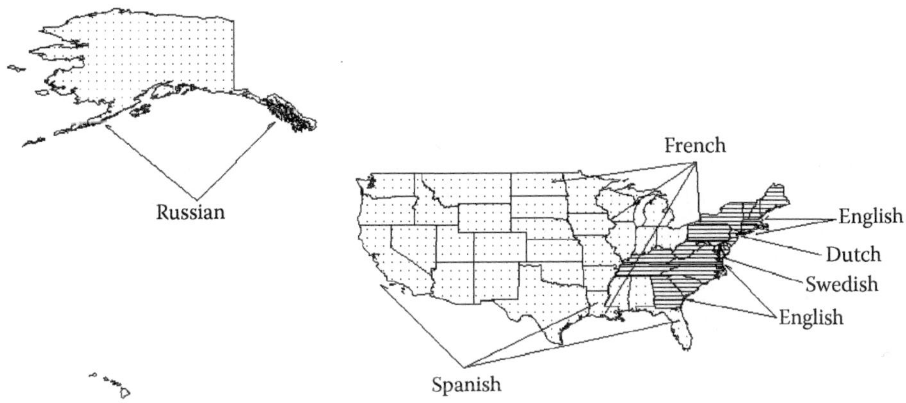

FIGURE 3.2
European influence in U.S. law.

influence, the *vara* is the standard unit of measurement of length used for land boundaries. There are several definitions of the *vara,* depending on where it was used. The same may be said for the French *arpent,* the English *rod,* and other units of land measurement. To assume one definition for any unit of measurement may result in an erroneous measurement.

In addition to geographical differences, historic differences must be taken into account when considering land tenure. Titles and boundaries were established under several different systems before the adoption of concepts in place at present time. Since the establishment of title is the very foundation of land tenure, and the establishment of boundaries locates and defines the extent of land tenure, it is of vital importance to fully understand the basic building block from whence it came. As aptly stated by the New York Court in *Dolphin Lane Associates, Ltd., v. Town of Southampton,* "the past must be explored to understand the present."

A significant retracement case involving early grants is that of *U.S. v. Champion Paper Company,* wherein the original grants were measured in *varas.* A later case involving earlier original grants is the Maryland case of *Ski Roundtop v. Wagerman,* wherein the two parties had identified an overlap of their respective boundaries. The Court of Appeals wrote that neither party could claim title to the area in question, since it was not a part of the original grant establishing either ownership. Citing a Maryland coal case, the court stated that a prerequisite for valid title is a continuous chain of title back to the sovereign, in this case as represented by the Calvert family. The opinion further stated that "no amount of successive transfers nor mere passage of time can metamorphose good title from void title."

The following seven sections illustrate the various bodies of law that have influenced current land law in the United States. Together, these collections of both common and civil law represent the great diversity of philosophies that are embedded within the current legal systems of the United States.

3.1.1 English Common Law

English common law began with the Magna Carta, first confirmed into law in 1225, as part of the centralizing of powers of the King during the middle ages after the Norman conquest. An important component of that document is the declaration of certain individual liberties, including the writ of habeas corpus, which prohibits a freeman being imprisoned or punished without the judgment of his peers. The common law coexisted with the civil law in England until the 17th century, when Parliament established a permanent check on the power of the English king and claimed the right to define the common law itself. That movement toward codification of the common law led to the writing of a key treatise on English common law, William Blackstone's *Commentaries on the Laws of England*, which still functions in American law as the definitive source for common law precedents prior to the existence of the United States.

The common law is a body of general rules prescribing social conduct enforced and applied by the court system. It evolves to meet changing social needs and improved understanding. The common law develops its principles from the grounds of decision in actual legal controversies. It is marked by its extensive use of the jury to provide the court with facts necessary for deciding a case. It is also marked by a doctrine of the supremacy of law (Hogue 1966).

Treatise writers and historians of the common law have long given custom a prominent place among the sources of this body of general rules. Maitland and Montague (1915) traced the evolution from customs originating in the "common wisdom and experience of society," through the stage of becoming "established customs," to the point at which they receive "judicial sanction in courts of last resort" (Hogue 1966).

Blackstone (1783) stated: "This unwritten, or common law, is properly distinguishable into three kinds: 1) General customs, which are the universal rule of the whole kingdom, and form the common law, in its stricter and more usual signification; 2) Particular customs, which for the most part, affect only the inhabitants of particular districts; and 3) Certain particular Laws, which, by custom, are adopted and used by some particular courts, of pretty general and extensive jurisdiction. All these are doctrines that are not set down in any written statute or ordinance, but depend merely upon immemorial usage, that is, upon common law, for their support."

At common law, possession is the root of all titles to land. A title rooted in possession entitles the possessor to an interest in the land. The kind of interest acquired depends on the nature of the interest available for acquisition at the time possession was taken. An interest acquired by possession, however, is not limited to the uses relied upon to establish the possession.

3.1.2 Roman Civil Law

Roman civil law is best illustrated by the Roman Civil Code known as the *Institutes of Justinian*, prepared under the direction of Emperor Justinian in

529 AD. That document serves as the root of today's civil law, including many of today's procedures for determining water boundaries. The following translations (Sandars 1874) of pertinent excerpts from that code illustrate the detail and the similarity of that code to modern water law.

Book II, Title I

> Section 1. By the law of nature these things are common to mankind—the air, running water, the sea, and consequently the shores of the sea. No one, therefore, is forbidden to approach the sea-shore, provided that he respects habitations, monuments, and buildings, which are not, like the sea, subject only to the law of nations... The sea-shore, that is the shore as far as the waves go at furthest, was considered to belong to all men.
>
> Section 2. All rivers and ports are public; hence the right of fishing in a port or in rivers, is common to all men.
>
> Section 3. The sea-shore extends to the limit reached by the greatest winter flood.
>
> Section 4. The public use of the banks of a river is part of the law of nations, just as is that of the river itself. All persons, therefore, are as much at liberty to bring their vessels to the bank, to fasten ropes to the trees growing there, and to place any part of their cargo there, as to navigate the river itself. But the banks of a river are the property of those whose land they adjoin; and consequently the trees growing on them are also the property of the same persons.
>
> Section 5. The public use of the sea-shore, too, is part of the law of nations, as is that of the sea itself; and therefore any person is at liberty to place on it a cottage, to which he may retreat, or to dry his nets there, and haul them from the sea; for the shores may be said to be the property of no man, but are subject to the same law as the sea itself, and the ground or sand beneath it.
>
> Section 20. Moreover, the alluvial soil added by a river to your land becomes yours by the law of nations. Alluvion is an imperceptible increase; and that is added by alluvion, which is added so gradually that no one can perceive how much is added at any one moment of time.
>
> Section 21. But if the violence of a river should bear away a portion of your land and unite it to that of your neighbour, it undoubtedly still continues yours. If, however, it remains for a long time united to your neighbour's land, and the trees, which it swept away with it, take root in his ground, those trees from that time become part of your neighbour's estate.
>
> Section 22. When an island is formed in the sea, which rarely happens, it is the property of the first occupant; for before the occupation it belongs to no one. But when an island is formed in a river, which frequently happens, then it occupies the middle of the river, it belongs respectively to those who possess the lands near the banks on each side of the river, in portion to the extent along the banks of each man's estate. But if the island is nearer to one side than the other, it belongs to those persons only who possess lands contiguous to the bank on that side. But

> if a river divides itself at a certain point, and lower down unites again, thus giving to any one's land the form of an island, the land still continues to belong to the person to whom it belonged before.
>
> Section 23. If a river, entirely forsaking its natural channel, begins to flow in another direction, the old bed of the river belongs to those who possess the lands adjoining its banks, in proportion to the extent along the banks of their respective estates. The new bed follows the condition of the river, that is, it becomes public. And, if after some time the river returns to its former channel, the new bed again becomes the property of those who possess the lands along its banks.
>
> Section 24. The case is quite different if any one's land is completely inundated; for the inundation does not alter the nature of the land, and therefore, if the water recedes, the land remains indisputably the property of the same owner.

As may be seen, the Roman code addresses many of the same issues relating to the ownership and boundaries of water bodies pertinent in today's society, including issues such as public ownership of navigable waters, the location of the boundary between publicly owned waters and privately owned littoral or riparian land, public access to navigable waters, riparian rights to newly formed lands, and ownership of newly formed islands (Robillard and Wilson 2014). Application of the Roman law to U.S. law regarding water boundaries is discussed more thoroughly in Section 7.3 of this book.

In addition to issues relating to water boundaries, the Roman civil code also addressed a wide range of issues relating to civil rights. These include ownership of property, familial relationships, inheritance rights, etc.

3.1.3 Spanish Civil Law

Most of the Spanish law influencing property rights in the United States today is derived from an early codification of civil law related to the Roman *Institutes of Justinian*. That collection of early Spanish law, entitled *Las Siete Partidas*, is a Castilian statutory code compiled in the 13th century during the reign of Alfonso X of Castile. As with the Roman *Institutes of Justinian*, *Las Siete Partidas* addressed a wide range of issues relating to civil rights. As implied by the title (*Siete Partidas*), this code is divided into seven parts, addressing various broad topics including natural law, public and military law, judicial organization, domestic relations, contracts, and criminal law. Several of the sections appear to have been based on the Roman *Institutes of Justinian*. The Spanish code is remarkably comprehensive, even covering topics such as who has the authority to hear confessions or perform baptisms and rules for tithing.

As an outpost of the kingdom of New Spain, the province of Texas shared with Mexico the aforementioned law of the parent sovereign, that of Castile.

This law remained essentially unchanged during Mexican Texas, which spanned the years 1821 to 1836. After the Texas Revolution, the Republic of Texas adopted the law of England. One of the exceptions was the law affecting land titles and certain water rights. Some of the elements of the Spanish legal system worked very well, and settlers were comfortable with it, since it was less complicated than English procedure. However, after the Civil War came an influx of common law lawyers and teachers trained in the Anglo-American system.

As a result of the continued application of the Spanish doctrine to land grants made before the Act of 1840, and those made thereafter, results in different rules may apply to adjacent tracts depending on the origin of the grant. For example, along the seashore, where the shore is flat, a common law grant may extend farther toward the ocean than an adjacent Spanish law grant due to the different system of measuring the coastal boundary. There is also a difference with regard to grants along rivers. Grants made under Hispanic rules extend only to the bank of a navigable waterway, as the sovereign was the owner of the bed of the river. Later common law grants extend to the center of a navigable waterway. The law for Spanish grants is also more restrictive regarding rights of irrigation, as Spanish grants do not contain such rights unless specifically granted.

Spanish law and its influence may also be found in the states of Arizona, New Mexico, California, and in southern Colorado, all of which were part of the Spanish territory, overseen and governed by the Spanish and Mexican governments. Land grants were made under those systems, establishing original grant boundaries, as well as influencing later subdivisions after the areas became part of the United States and the several states.

As with the Roman *Institutes of Justinian*, the primary impact of *Las Siete Partidas* on contemporary property rights in the United States is in dealing with the ownership and boundaries of waters. Translations of portions of the code addressing the boundary of the public owned coastal waters are as follows (Scott 1931):

> The things which belong in common to the creatures of this world are the following, namely; their, the rain-water, and the sea and its shores, for every living creature can use each of these things, according as it has need of them. For this reason every man can use the sea and its shore for fishing or for navigation, and for doing everything there which he thinks may be to his advantage.
> …and all that ground is designated the shore of the sea which is covered with the water of the later at high tide during the whole year, whether in winter or in summer.

Thus, the Spanish Civil Law appears to continue to call for both the sea and the seashore to be publicly owned. Yet, it appears to depart from the Roman Civil Law in that it calls for the consideration of the tidal cycles as opposed to seasonal high waters as with the Roman code. See Section 6.3 of this book for a complete discussion of this topic.

3.1.4 French Civil Law

Louisiana private law is primarily a Napoleonic system and is the only U.S. state partially based on French and Spanish codes and, ultimately, Roman law. In Louisiana, private law was codified into the Louisiana Civil Code; however, the law has considerably merged with American law, especially in the public law and in the judicial system. In fact, any innovation, whether private or public, has been decidedly common law in origin.

French civil law is based on the Napoleonic Code of 1804, named after the emperor. That Code consists of three sections: the law of persons, property law, and commercial law. Rather than a collection of statutes or listing of case law, that code sets out general principles as rules of law. Sections of the French Civil Law, including laws regarding property, contacting, and business entities, were strongly influenced by the Roman Civil Law.

The influence of the French Civil Law is especially noticeable in the state of Louisiana. Actually, the first Louisiana Civil Code Digest of 1808 was written in French, although then translated into English. Therefore, significant differences exist between Louisianan civil law and the common law found in most other American states. Nevertheless, Louisiana law does have many of the terms and concepts unique to American law, including *forced heirship, redhibition*, and *lesion beyond moiety*.

In Quebec, whose private law is also primarily of French civilian origin, development has been along the same lines, having adapted in the same way as Louisiana to the public law and judicial system of Canadian common law. By contrast, Quebec private law has innovated mainly from civilian sources. To a lesser extent, other states formerly part of the Spanish Empire, such as Texas and California, have also retained aspects of Spanish civil law into their legal system, for example, community property. The legal system of Puerto Rico exhibits the same inclinations that Louisiana has shown, that is, the application of a civil code whose interpretations rely on both the civil and common law systems. Because Puerto Rico's Civil Code is based on the Spanish Civil Code of 1889, available jurisprudence has tended to rely on common law innovations due to the code's age and, in many cases, obsolete nature.

While the French settled in several areas, including Maine, northern New Hampshire and Vermont, Michigan, Ohio, and especially along the Mississippi River, French law did not develop fully, nor was it widely applied. The French influence did, however, affect the establishment of titles and their boundaries and therefore had an impact on tenure.

In Massachusetts, the Supreme Court adopted from the French Civil Code the rule as to riparian rights, which is now found throughout the common law world. Ohio took over from the civil law its doctrine as to cases regarding agency relationships.

3.1.5 Russian Civil Law

Russian America was the name of the Russian colonial possessions in the Americas from 1733 to 1867. Several settlements crossed parts of what are now the U.S. states of California, Alaska, and two ports in Hawaii. Formal incorporation of the possessions by Russia did not take place until the Ukase of 1799, which established a monopoly for the Russian-American Company and also granted the Russian Orthodox Church certain rights in the new possessions. Many of its possessions were abandoned in the 19th century. In 1867, Russia sold its last remaining possessions to the United States.

With grants, the chief administrator of the Russian-American Company would assign a plot of land for housekeeping and fishing, taking special care only "to avoid contestations between the settlers and the natives." These assignments were not accompanied by any written titles or other legal documents. Rather, "the first occupation and using of a certain locality, whether by an individual or by a community, notwithstanding the lack of formalities, conferred unquestionable right of possession." No one, not the Russians nor the Alaska natives, had anything more according to Russian law.

The provisions of Russia's 1844 Charter remained in force until the time of cession to the United States in 1867. They constituted a part of the laws in force in the territory subsequently acquired by the United States and are therefore subject to judicial notice by the American courts. The opinion of Judge Wickersham in *United States v. Berrigan*, 2 Alaska 446, states as follows:

> Where territory has been acquired by the United States from a foreign power, its courts will take judicial notice of the laws which prevailed there up to the time of such acquisition. They are not, as to such acquired territories, foreign laws, but laws of an antecedent government. (*United States v. Perot*, 98 U.S. 428, 25 L. ed. 251; *Fremont v. United States*, 17 How. (U.S.) 542, 15 L. ed. 241)

3.1.6 Dutch Civil Law

The first constitution of New York in 1777 prohibited interference by the state with "any grants of land within the state, made by the authority of the king or his predecessors prior to 1775." Therefore, the quality and extent of "any grants of land within the state made by the authority of said king" must be determined by the rules of law that operated upon the grant at the time the grant was made and title passed under it (N.Y. Const. of 1777, § 36).

Prior to the English taking over what is now Pennsylvania and the Delaware, the Swedes and the Dutch each in turn set up a government. Considered under Dutch rule, the first period extended from 1609 to 1638; the Swedish rule extended from 1638 to 1655; and the Dutch rule, the second period, extended from 1655 to 1664. The Swedish settlement was governed solely under Swedish law, but when the Dutch gained control of the Delaware, this was placed under the authority of the director and council of New Amsterdam. The Dutch organized local governments and set up courts, which seem to have been required to oversee only rather trivial matters.

For a time after authority was transferred, the Delaware lands were allowed to maintain the authority of their Dutch courts, but in 1676, the district was called upon to enforce the code of laws adopted in 1664 at Hemsted on Long Island. This is known as the Book of Laws of the Duke of York. The preamble to this code states that the measures contained within it were "Collected out of the several laws now in force in his Majesties American Colonyes and Plantations." But it was the laws of Massachusetts and New Haven especially that influence Gov. Nicholls and his associates in the preparation of the code of 1664.

With the coming of William Penn, a new era began in the legal system in that region. Out of his own bitter experiences with the English law came certain conceptions of justice that he hoped would be adopted in the new world. While still in England, he had drawn up a fundamental body of laws, and the Great Law of Chester adopted in 1682 contained these juristic ideals.

The Quakers showed great reluctance in the earlier period to applying the harsh sanctions set up by the mother country and so occupied themselves with law making and otherwise set themselves against the application of the laws of Great Britain, declaring through their superior court that English statute law could not be applied. This sentiment gradually passed away, and in the next century, by the Act of 1718, a large body of English law was expressly made applicable to Pennsylvania, "because the statutes of Parliament did not extend to this province" (Gipson 1915).

3.1.7 Aboriginal Law

Aboriginal land law is the common law doctrine holding that indigenous people have customary land tenure rights over a territory even after the assumption of sovereignty by a colonizing power. The primary basis for such indigenous land rights is continuous and exclusive occupancy of land over an extended period. In addition, additional basis for many aboriginal claims in the United States are the action of governments which antedate the current government (e.g., a grant from Great Britain and Spain), treaties with the United States, acts of Congress, executive actions, or purchase (Morisset 2001).

Aboriginal law as it relates to aboriginal title derives much from aged international law in part not only because of the tendency of contemporary judges to construct it accordingly but also because almost all of these claims have a factual background traceable back a century or more. *Aboriginal title* has been defined as a species of land ownership that is unique to aboriginal law and that, as the present flow of the law implies, comes with a very unique set of conditions for it to be recognized and enforceable, very different from the law of *freehold*. It is also known as indigenous title, native title (particularly in Australia), original native title (particularly in the United States), and customary title (particularly in New Zealand). Aboriginal title jurisprudence is related to indigenous rights, influencing and influenced by nonland issues, such as whether the government owes a fiduciary duty to indigenous peoples. While the judge-made doctrine arises from customary international law, it has been codified nationally by legislation, treaties, and constitutions.

An important concept in understanding such aboriginal land rights is that indigenous people's relationship with land does not easily adapt to western concepts of real property and economics. Rather, it is intrinsically connected to cultural, physical, and spiritual well-being in addition to more obvious economic benefits. Native Americans historically did not have a concept of private ownership of land but instead followed a system of tenure of land for the perpetual use and occupancy of the respective tribe. Consequently, Indian and non-Indian cultures held significantly divergent views concerning their uses of and relationships with the land. Where Europeans had an extremely well-defined body of property law based on clear ideas of individual ownership, Native American tribes generally held land communally with shared benefits and burdens. Even when European settlers and developers acquired lands from the Indians by trade and by purchase, each saw the result in a different light. The purchasers, based on their knowledge and understanding of land ownership, believed they were acquiring a title to the land in fee simple. Native Americans, based on their beliefs, believed that they were allowing purchasers the right of use and occupancy of the lands and welcomed that shared use. Without having same concepts and beliefs of land ownership, it was not possible to agree on the ultimate result. This obviously led to significant problems and disagreements.

The colonizing European nations developed the Doctrine of Discovery to justify the process of dominion and subsequent settlement. That doctrine has been described as holding that newly arrived Europeans "immediately and automatically acquired legally recognized property rights in native lands and also gained governmental, political, and commercial rights over the inhabitants without the knowledge or the consent of the Indigenous peoples." The discovery doctrine remains a foundational legal principle in the United States, New Zealand, Australia, and Canada, as well as elsewhere.

European views of aboriginal land rights were by no means uniform. English, French, Spanish, Dutch, and Swedish concepts varied considerably in their emphasis on discovery, papal authority, royal grant, feudal right, possession, and purchase. The English believed that, based on feudal tenure concepts, the fundamental principle is that the king was the original proprietor and the only possible source of title. The English charters, with the exception of Rhode Island, denied the property rights of the Indians. The Pennsylvania charter of 1681 gave William Penn full rights in his territory, including rights of assigning and granting. However, some Puritans, along with William Penn, were careful to buy from the Native Americans the lands they wished to occupy. Even so, as described previously, the Native Americans believed they were assigning, or conveying, only the right of shared use in the lands. Concerning the French, there are but few examples of grants of Native American lands to Frenchmen, either in public or private capacity. The Dutch based their claims on purchase rather than discovery, or royal or papal patents. They believed that right of discovery of that held by another was wrong. Consequently, records of the Dutch West India Company show that their patents to Dutch settlers transferred ultimate right of ownership only after the grantee had purchased from the local Native American tribes. The policy of the Swedes was similar to that of the Dutch. The Spanish, on the other hand, did not see the same importance as other European powers concerning land ownership. Spaniards thus acquired land from the Native Americans by purchase, royal grant, and forced seizure. Natives could, and did, sell land to private individuals.

The United States was the first colonial nation to recognize aboriginal land rights with early opinions, such as *Johnson v. M'Intosh*. Such opinions were based on the concept that Native Americans did not have full title to the land because they did not have a concept of individual property rights. Rather, because of the discovery and conquest of this previously unclaimed continent (*terra nullis* or vacant land) by Europeans and the subsequent revolution and treaties, the United States had exclusive title to all land not previously conveyed by the European sovereigns. Yet, under such policy, Native Americans might have the right of possession of the land on which they live as *usufruct* (the right to enjoy something owned by someone else), but only at the pleasure of the United States.

Beginning in 1830, the U.S. Indian Removal Acts formalized a policy that resulted in almost complete extinction of aboriginal land interests in the eastern states. Passage of that law in 1830 came after considerable debate and a close vote in Congress, with notables such as Abraham Lincoln and Davy Crockett arguing against the law. Basically, the law provided for Native American tribes to cede their land in the east for payment and land in the west. Some tribes, such as the Seminoles, did not leave peaceably, resulting in the three Seminole Wars. Subsequently, a series of legislation and treaties continued to alienate Native American land and designate the boundaries of reservations for the various tribes.

Currently, there reportedly are 310 Native American reservations in the United States (Figure 3.3), although there are 550 recognized tribes, with a total acreage of 55,700,000 acres or 2.3% of the area of the United States. Thus, Native American tribes now do have sovereign rights over the land on reservations that have been granted through treaties with the United States.

An important concept of Indian Tribal rights and the duty of the United States to enforce them is the concept of trust responsibility, which was first expressed in a decision by Chief Justice John Marshall in *Cherokee Nation v. Georgia*.

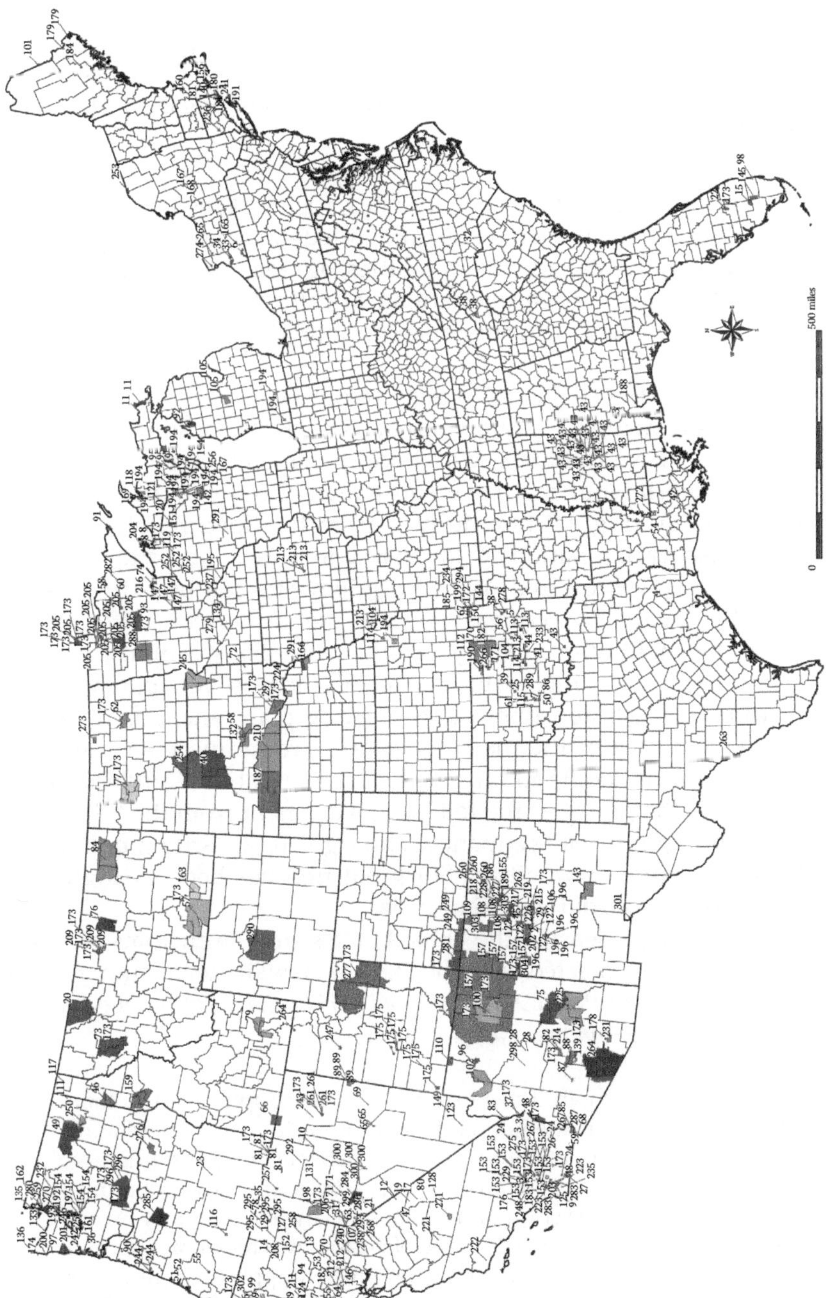

FIGURE 3.3
Indian reservations in the continental United States. (Courtesy of the U.S. Bureau of Indian Affairs and the National Park Service.)

> In Cherokee Nation, the tribe filed an original action in the Supreme Court to enjoin enforcement of state laws. The Court refused original jurisdiction, finding that the tribe was not a state of the United States nor a foreign state, and thus not entitled to bring suit directly before the Court. The Court concluded rather that tribes were "domestic dependent nations" and that their relationship to the United States resembled "that of a ward to his guardian."

Based on that decision and subsequent ones, the trust responsibility doctrine has developed. That doctrine has guided federal action toward Native Americans in governmental actions in treaties, agreements, statutes, executive orders, and administrative regulations, and the trust relationship has become one of the primary cornerstones of aboriginal law in the United States.

In *New York v. Shinnecock*, Justice Bianco wrote extensively on the law of aboriginal title, borrowing from many previous cases: "Aboriginal title refers to the Indians' exclusive right to use and occupy lands they have inhabited from time immemorial, but that have subsequently become discovered by European settlers. Aboriginal title derived from the doctrine of discovery provided that discovering nations held fee title to these lands, subject to the Indians' right of occupancy, and use. Aboriginal title can be extinguished by treaty, by the sword, by purchase, by the exercise of complete dominion adverse to the right of occupancy, or otherwise. For extinguishment to occur, the sovereign must intend to revoke the Indians' occupancy rights. The intent to extinguish aboriginal title must be plain and unambiguous based on either the face of the instrument or surrounding circumstances. Extinguishment cannot be lightly implied. The foundational underpinning of this standard is the policy of the federal government from the beginning to respect the Indian right of occupancy. Thus, given this strong policy, any ambiguity on the issue of whether aboriginal title has been extinguished must be resolved in favor of the Indian tribe."

Practically nowhere, at least in the United States, are the customs of a people so entrenched and protected as in the State of Hawaii. Regarding this, the court noted in the case of *Public Access Shoreline Hawaii by Rothstein v. Hawai'i County Planning Commission by Fujimoto*:

> We perceive the Hawaiian usage exception to the adoption of the common law to represent an attempt on the part of the framers of the statute to avoid results inappropriate to the isles' inhabitants by permitting the continuance of native understandings and practices which did not unreasonably interfere with the spirit of the common law. The statutory exception is thus akin to the English doctrine of custom whereby practices and privileges unique to particular districts continued to apply to the residents of those districts even though in contravention of the common law. This is not to say that we find that all the requisite elements of the doctrine of custom were necessarily incorporated in § 1-1. Rather we believe that the

> retention of a Hawaiian tradition should in each case be determined by balancing the respective interests and harm once it is established that the application of the custom has continued in a particular area.

Aboriginal title may not be alienated, except to the federal government or with the approval of Congress. It must be stressed that Aboriginal title is distinct from the lands that Native Americans own in fee simple and occupy under federal trust.

In recent times, the Alaska Native Claims Settlement Act (1971) extinguished all aboriginal title in Alaska, although the legitimacy of the act remains disputed by some Alaskan natives. Indian Land Claims Settlements extinguished all aboriginal title in Rhode Island (1978) and Maine (1980).

Most other colonized nations have also struggled with aboriginal land rights. As an example, Australia began to experience native land rights litigation in the 1970s, when indigenous Australians became more politically active. After litigation (e.g., *Milirrpum v. Nabalco Pty Ltd.*) on behalf of Aborigines claiming all of the country, the Aboriginal Land Rights Act 1976 returned approximately 40% of the Northern Territory to Aboriginal ownership. Another Land Rights Act enacted in 1981 had a similar effect in South Australia.

Based on those acts and others, current indigenous land holdings have been estimated to currently include about 15% of the Australian continent. Those land holdings are available for the use, benefit, and residence of both Aboriginal and Torres Strait Islander people. While some holdings are alienable, leasehold, or Crown reserve, much are inalienable freehold land, which cannot be sold or mortgaged. While that restriction is often considered a barrier to business development, it does ensure continuation of indigenous ownership for future generations.

The Canadian experience with aboriginal land rights has been somewhat different from that of both the United States and Australia. The Canadian policy began with a 1763 English royal proclamation. While that proclamation claimed ownership of what is now Canada by the English King, it recognized that indigenous peoples already on the lands had a prior claim, not extinguished by the arrival of the Europeans, and the Crown had a fiduciary duty to protect it for the Aboriginal peoples.

That policy recognizes Aboriginal rights as *sui generis*, or unique collective rights based on the land being the ancestral territory of indigenous groups. Aboriginal title cannot be held by individual aboriginal persons; it is a collective right to land held by all members of an aboriginal nation. Decisions with respect to that land are also made by that community. Such rights are not granted from an external power but are the result of the occupation of and relationship with their territory as well as their social customs and political systems. In an early judicial ruling to protect such rights, the 1763 Royal Proclamation also stated that such Aboriginal rights could only be extinguished by treaty with the Crown and not ceded directly to settlers.

Subsequent to the 1763 proclamation, Crown representatives and leaders of Aboriginal groups executed treaties throughout most of Canada in efforts to resolve issues of land rights. Those treaties typically traded tenure to land for annuities and certain privileges and legal exemptions. Not all groups signed such treaties, and groups that have never signed treaties can still lodge land claims against the government. As an example, in most of British Columbia, Aboriginal title has never been transferred to the Crown. The British Columbia Treaty Commission defines Aboriginal Rights, which are separate from Treaty Rights, as the practices, customs, and traditions unique to First Nations that were participated in prior to contact with Europeans. Aboriginal Rights, such as the right to hunt and fish, are constitutionally protected and cannot be extinguished by any government. In addition, a number of Indian reserves have been established in Canada where the legal title is held by the federal government for the use and benefit of an Indian group. Currently, there are over 2700 such reserves covering 3,123,550.8 hectares (7,715,170 acres). Also, in the new territory of Nunavut ("our land"), which includes the Arctic Islands and which was carved out of the North West Territories in 1999, title to about 350,000 square kilometers (of the total area of Nunavut of 1.9 million square kilometers) was returned to the Inuit.

As may be seen from the previous discussion, Aboriginal land law generally holds that indigenous people have customary land tenure rights over a territory even after the assumption of sovereignty by a colonizing power. Those rights were not granted by formal title, rather they are the result of the occupation of and relationship with their territory as well as their prevailing social customs. The recognition of such rights has varied widely among colonizing nations and has evolved considerably with time and social mores. Most aboriginal people see their rights as being inherent, collective, and well defined, encompassing such areas as land ownership, education, and the right to self-government.

The overall concept of Aboriginal Title is well summarized by the following 1997 quote from Justice La Forest, then Chief Justice of the Canadian Supreme Court: "In my view, the foundation of aboriginal title (is) the fact that when the settlers came, the Indians were there, organized in societies and occupying the land as their forefathers had done for centuries. This is what Indian title means. Aboriginal peoples have a legal right to occupy and possess certain lands, the ultimate title to which is in the Crown. As well, the inescapable conclusion from the Court analysis of Indian title up to this point is that the Indian interest in land is truly sui generis. It is more than the right to enjoyment and occupancy although it is difficult to describe what more in traditional property law terminology."

3.1.8 Tribal Law

As distinguished from aboriginal law, tribal law is that associated with the right of self-governance and self-sufficiency typically maintained by Native

American tribes. Tribal law recognizes the inherent sovereignty or authority of indigenous tribes to govern themselves within the borders of the United States. The U.S. federal government recognizes tribal nations as "domestic dependent nations" and has established a number of laws attempting to clarify the relationship between the federal, states, and tribal governments. While Native American tribal sovereignty is partially limited as "domestic dependent nations," the same may be stated about the sovereignty of the federal government and the individual states, each of which is limited by the other. Therefore, tribal sovereignty is another form of parallel sovereignty within the U.S. constitutional framework, constrained by, but not subordinate to, other sovereign entities.

Such rights can affect not only Native Americans but also non-Native Americans and property outside of reservations held by others. Because of the supremacy of Native American treaties under the U.S. Constitution, tribal law may be superior to rights based on state or common law (Morisset 2001). Since treaties with Native Americans have their roots in the U.S. constitution, tribal law has its basis in that document and is usually considered to preempt state law. Further, since the United States maintains fee title to much Native American property, that fact also dictates federal law preeminence. Tribal sovereignty is dependent on, and subordinate to, only the federal government, not states, under *Washington v. Confederated Tribes of Colville Indian Reservation* (1980), and tribes are sovereign over tribal members and tribal land under *United States v. Mazurie* (1975). It should be noted, however, that the United States retains control over the scope of tribal law making and that laws adopted by Native American governments must also pass the Secretarial Review of the Department of Interior.

A large component of cases dealing with tribal law involves habitat and property. As an example, it has been held in some cases that Native American reserves extend to and include adjacent water, submerged land, and tidelands (e.g., *Alaska Pacific Fisheries v. United States, State v. Edwards*). Even where shorelines are not within or adjacent to a Native American reservation, the state still may be required to regulate in accordance with tribal rights. Tribal law also applies to hunting and fishing rights. Generally, Native Americans have exclusive rights to hunt and fish on their reservations. In addition, those rights may extend to off-reservation sites, which are their usual and accustomed locations for such activities, even extending to areas where the underlying submerged lands or tidelands are in private ownership in some cases (*U.S. v. Washington*). In addition, it has been held in some cases that the treaty right to hunt or fish includes implied rights to have the fishery resource protected by appropriate preservation of habitat or by proper design and construction of culverts and roads. Similar rights have also been held to include on and off-reservation water rights to consumptive uses of groundwater and surface waters in some instances.

Tribal government regulatory powers have been held to include health, zoning, and land use regulations on reservations, including civil jurisdiction over non-Native Americans and fee lands not held by Native Americans. A 1981 case, *Montana v. U.S.*, ruled that Native American tribes have inherent power over their internal affairs as well as civil authority over nonmembers as necessary for the protection of the political integrity, the economic security, or the health and welfare of the tribes.

3.2 Origins of Land Tenure in the Original Colonies

In the United States, the origins of most currently recognized title to land in the areas covered by the original colonies are royal grants from the sovereign of the several European nations claiming land based on "discovery." By almost any standard, the basis of the claims should be more appropriately considered as a conquest rather than a discovery since there were reportedly more than 800,000 Native Americans in 330 tribes living in what is now the United States at the time of the European "discovery." Furthermore, the Native Americans were certainly aware of rights to land based on the following comment in report sent to the British Lords of Trade in 1714: "Each Indian nation is perfectly well acquainted with its exact bounds; the same is again divided into due proportions for each tribe and afterwards subdivided into shares to each family, with all which they are most particularly acquainted. Neither do they ever infringe upon one another or invade their neighbors' hunting grounds." Nevertheless, the Europeans did not recognize the ownership rights of the Native Americans. Some settlers made token "purchases" for small sums, but those transactions were only to pacify the Native Americans or to give the settlers grounds for the defense of their claims against other settlers. The U.S. Supreme Court has continued this policy in its decisions, which have recognized only those Native American land rights that have been confirmed by the government of the United States by treaty or congressional actions. Regardless of the fairness of the acquisition by the Europeans, it has withstood the test of time, and as a result, current title to land in the area covered by the original colonies has as its basis, royal grants from European sovereigns.

Although there were overlapping European claims, the dominant claim was that of King James I of England for all of the land in America between the French settlement on the St. Lawrence and the Spanish settlement in Florida. That claim was based on an exploration cruise in 1498 by John Cabot along that coastline when he went as far south as Cape Hatteras.

Typical of the royal grants to the original colonies was the first British grant in 1606 from King James I to Sir Thomas Gates and three other British noblemen for an area lying 50 miles south and 50 miles north from the seat of the first location in America (Jamestown, Virginia) and directly into the mainland for 100 miles, together with all islands within 100 miles of the

FIGURE 3.4
Limits of 1606 Jamestown Grant (land within 100 miles of the coastline between 34° and 41° north latitude).

coastline between 34° and 41° north latitude (Figure 3.4). Each member of the Jamestown colony was promised one share of stock in the company and title to 100 acres of land at the end of an initial 7-year period during which all resources were to be shared.

Following the initial grant, there were numerous additional British royal grants, many of them overlapping, as well as changes to the initial grant. The next permanent British colony after Jamestown was the settlement of the pilgrims on the Mayflower at Plymouth in 1620. Those immigrants purchased land from the Virginia Colony prior to departure from England. Due to either storms en route or poor navigation, they landed to the north of the limits of that grant near Plymouth Rock in Cape Cod Bay. There, they found an abandoned Native American village with remnant cornfields and began their settlement. Seven years later, they were able to purchase that land that they occupied from a company that had obtained a royal grant to that area.

Following the Plymouth colony, the number of British colonies in the New England area multiplied rapidly, with colonies being established in Maine, Salem (Massachusetts), Connecticut, Rhode Island, New Hampshire, and Vermont. In addition, in 1663, the British took, by force, the Dutch colony in New York after years of disputes, as well as a former Swedish colony along the Delaware River. The original Dutch grants preceding the takeover were confirmed by the British when they took possession. During the same period, other colonies developed, including the current state of Pennsylvania based on a grant to William Penn, Maryland based on a grant to George Calvert (Lord Baltimore), to New Jersey based on a grant to Berkeley and Carteret, as well as grants covering many of the trans-Appalachian regions.

In another grant during that period, on March 24, 1663, Charles II issued a new charter to a group of eight English noblemen, granting them the land of Carolina, as a reward for their faithful support of his efforts to regain the throne of England. The eight were called "Lords Proprietors" or simply "Proprietors." That charter granted the Proprietors title to all land between the southern boundary of the Virginia Colony (36° north latitude) to 31° north latitude, along the coast of present-day Georgia. In 1665, the charter was revised to extend the northern boundary to 36° 30 minutes north to include lands of settlers along the Albemarle Sound, who had left the Virginia Colony. That revision also extended the southern boundary southerly to 29° north, which is just south of Daytona Beach, Florida. That revision had the effect of including an existing Spanish settlement in St. Augustine, Florida. The charter granted all of the lands between those bounds from the Atlantic Ocean westerly to the Pacific Ocean.

ROYAL CHARTER

George the second, by the grace of God, of Great Britain, France and Ireland, king, defender of the faith, and so forth. To all to whom these presents shall come, greeting.

Whereas we are credibly informed, that many of our poor subjects are, through misfortunes and want of employment, reduced to great necessity, insomuch as by their labor they are not able to provide a maintenance for themselves and families; and if they had means to defray their charges of passage, and other expences, incident to new settlements, they would be glad to settle in any of our provinces in America where by cultivating the lands, at present waste and desolate, they might not only gain a comfortable subsistence for themselves and families, but also strengthen our colonies and increase the trade, navigation and wealth of these our realms....

And whereas we think it highly becoming our crown and royal dignity, to protect all our loving subjects, be they ever so distant from us; to extend our fatherly compassion even to the meanest and most unfortunate of our people, and to relieve the wants of our above mentioned poor subjects; and that it will be highly conducive for accomplishing those ends, that a regular colony of the said poor people be settled and established in the southern territories of Carolina....

Know ye, therefore that we greatly desiring the happy success of the said corporation, for their further encouragement in accomplishing so excellent a work have of our aforesaid grace, certain knowledge and mere motion, given and granted by these presents, for us, our heirs and successors, do give and grant to the said corporation and their successors under the reservation, limitation and declaration, hereafter expressed, seven undivided parts. The whole in eight equal parts to be divided, of all those lands, countrys and territories, situate, lying and being in that part of South-Carolina, in America, lying and being in that part of South Carolina, in America, which lies from the most northern part of a stream or river there, commonly called the Savannah, all along the sea coast to the southward, unto the most southern stream of a certain other great water or river called the Altamaha, and westerly from the heads of the said rivers respectively, in direct lines to the south seas; and all that share, circuit and precinct of land, within the said boundaries, with the islands on the sea, lying opposite to the eastern coast of the said lands, within twenty leagues of the same, which are not inhabited already, or settled by any authority derived from the crown of Great Britain....

All which lands, countries, territories and premises, hereby granted or mentioned, and intended to be granted, we do by these presents, make, erect and create one independent and separate province, by the name of Georgia, by which name we will, the same henceforth be called.

FIGURE 3.5
Excerpts from the Royal Grant of 1732 for the Colony of Georgia.

All of these were based on British royal grants to individuals or companies, who were essentially land speculators with the primary objective of financial gain. An exception to this was one of the later colonies in the current state of Georgia. King George II bought back a portion of the Carolinas and granted the Georgia area to General James Oglethorpe in 1732 for the purpose of establishing a colony for people of respectable families who were confined in prisons as insolvent debtors (Chandler 1945). That philosophy is reflected in excerpts from the royal grant for the colony (Figure 3.5).

As may be seen in the description contained in the charter, the granted land was described as lying between the Savannah and Altamaha rivers, up to their respective headwaters and then extending westerly "to the south seas." At the end of the French and Indian War, a British royal proclamation in 1763 fixed the northern border of Florida. Three months later, Georgia's southern boundary was further defined as the north boundary of Florida, and its western boundary was fixed at the Mississippi River, including portions of what is now Alabama and Mississippi (Figure 3.6). The area between the Altamaha and St. Marys rivers was annexed to Georgia by the same proclamation.

In all of the areas covered by the original colonies, the origins of most currently recognized title to land in the areas are royal grants from the sovereign of the European nations. Those large royal grants were typically subdivided into smaller land tracts by the grantees and sold to the original settlers.

In the New England colonies, the settlement was typically planned prior to distribution of the land. The land was usually subdivided by survey, with lots grouped along roads and around a village square. Formal subdivision plats of each settlement were typically prepared, which allowed land to be

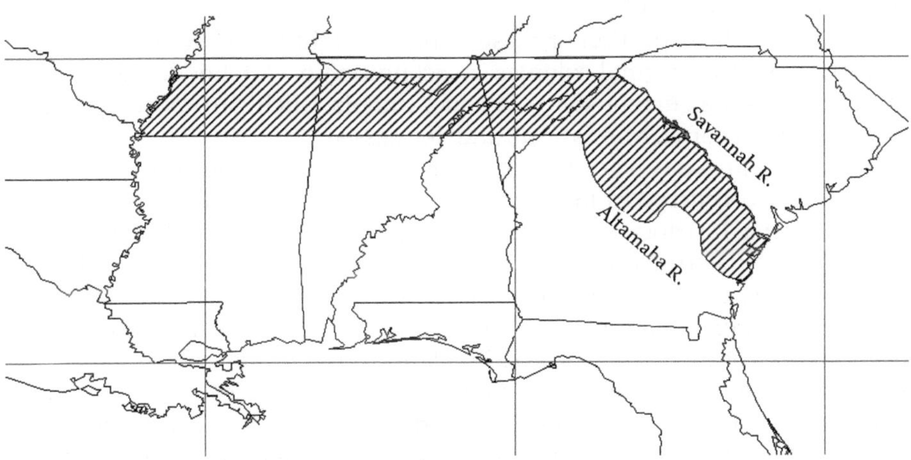

FIGURE 3.6
1732 Royal grant for the Georgia colony (land lying between the Savannah and Altamaha rivers, up to their headwaters and then extending "to the south seas").

described in reference to the plats. As the land was conveyed to the individual settlers, land records were carefully recorded. As a result, settlement patterns were relatively regular, with few gaps left between lots within each township or the townships themselves (Linklater 2002).

In other colonies, differing approaches were used. For example, in Virginia, settlers were allowed to choose their parcel of land and then have it surveyed and registered. Thus, highly irregular settlement patterns resulted, with gaps left between the more desirable tracts. In the Carolinas and Georgia, the proprietors of the original sovereign grants planned to presurvey their colonies into rectangles, with lines run in cardinal directions. Despite such plans, the urgent demand for land resulted in a complete breakdown of those plans. Newly arriving settlers were given "head rights" that entitled them to a certain quantity of land at a location of their choosing as long as it was not occupied. Settlers were supposed to have their land surveyed and the title recorded after they selected the land, but often, these processes were not accomplished. As a result, an irregular, scattered pattern of settlement developed with frequent land disputes (Clawson 1968).

3.3 Origins of Land Tenure in Public Domain Lands

Unlike the land within the original 13 colonies, most of the remainder of the land currently within the United States was acquired after the colonial period and passed though formal ownership of the United States. Thus, over three quarters of the current land of the nation was once considered public domain land (Figure 3.7).

Cessions Land—A portion of these public domain lands were acquired by cessions from the original colonies as a condition of statehood. The cessions land includes the current states of Ohio, Illinois, Michigan, Wisconsin, and portions of Alabama, Mississippi, and Tennessee.

Louisiana Purchase—The most dramatic increase in the territory of the United States occurred in 1803, when President Thomas Jefferson purchased the Louisiana Territory from France. That purchase involved over 800,000 square miles covering all or part of 13 states, including Arkansas, Colorado, Iowa, Kansas, Louisiana, Minnesota, Missouri, Montana, Nebraska, North Dakota, Oklahoma, South Dakota, and Wyoming. The purchase almost doubled the size of the United States. The land in the purchase was especially strategic for the United States in that it included New Orleans and its control of navigation on the Mississippi River and, thus, access to much of the inland areas of the continent. The purchase also had significance in international politics. It almost totally removed France from the Americas and reduced the likelihood of any other European power getting a foothold in North America

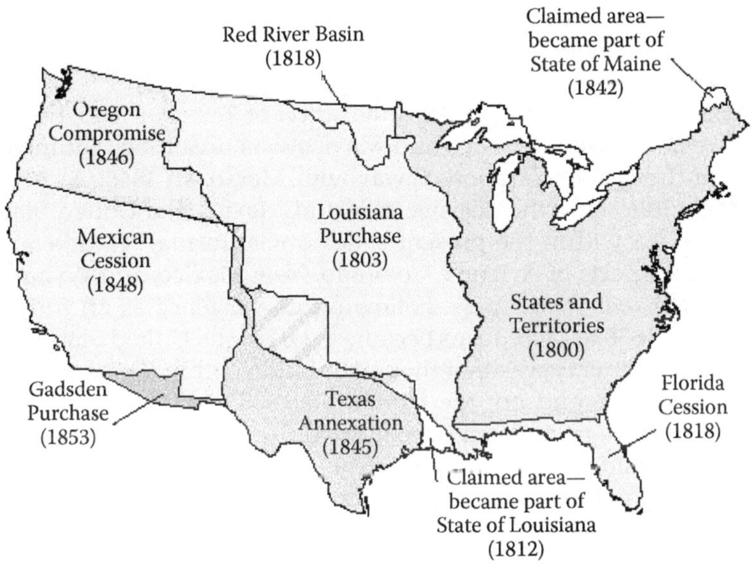

FIGURE 3.7
Land acquisitions of the United States, 1800–1860. (Courtesy of the U.S. Geological Survey.)

(Clawson 1964). France agreed to the purchase due to that nation's sore need for money caused by its involvement with European wars. Furthermore, with its attention required in Europe, France feared that it would not be able to defend and hold its claims in North America.

Red River Basin—To resolve territorial issues remaining after the Treaty of Ghent, which had ended the war of 1812, a new treaty was executed with Great Britain in 1818 that clarified the northern boundary of the Louisiana purchase. The additional land acquired under the new treaty was called the Red River Basin. That land later became parts of the states of Minnesota and North Dakota.

Florida Cession—In 1819, just 16 years after the Louisiana Purchase, the United States purchased, from Spain, the area within the current state of Florida and portions of Alabama and Mississippi. Spain agreed to the sale for some of the same considerations as those associated with France and the Louisiana Purchase. Spain had not been especially successful in settling the area and feared eventual loss of the territory to the aggressive new nation based on several border incidents.

Oregon Compromise—The area within the present states of Idaho, Oregon, and Washington and parts of Montana and Wyoming was of interest to both the United States and Britain. For the period between 1818 and 1844, the territory was the subject of a joint settlement and control treaty between the nations. That arrangement proved unsatisfactory to the settlers from the

United States due to the dominance of Britain's Hudson's Bay Company. As a result, the United States pressed for, and in 1845 obtained, a treaty providing for sole ownership of the territory by the United States.

Mexican Cession—Following the annexation of Texas in 1845, the relationship between the United States and Mexico was unsettled. Ultimately, this resulted in the U.S. declaration of war with Mexico in 1846. As part of the peace treaty following the decisive defeat of Mexico, the United States purchased the area within the present states of California, Nevada, and Utah and significant parts of Arizona, Colorado, New Mexico, and Wyoming.

Purchase of Lands from Texas—Following its existence as an independent republic from 1836 to 1845, Texas became a part of the United States by treaty. One term of that treaty was that the public lands within the new state would belong to the State and not the United States. Thus, the lands within the State of Texas have never been part of the U.S. public domain. Nevertheless, in 1850, the United States purchased, from Texas, its claim to almost 79 million acres lying within the present states of Colorado, Kansas, New Mexico, Oklahoma, and Wyoming.

Gadsden Purchase—In 1853, the United States purchased, from Mexico, a strip of land that had been the subject of a continued dispute. That land lies across southern Arizona and New Mexico.

Alaska—The last great territorial acquisition of the United States was the purchase, from Russia in 1867, of what is now the State of Alaska. Although the Russians had long claimed that territory, their primary interest had been in furs. As a result, there had been limited settlement attempts. Therefore, when the sea otter became nearly extinct and other fur species depleted, Russia was agreeable to the sale of the territory (Clawson 1964).

With the public domain lands (Figure 3.8), it was in the interest of the United States to encourage settlement in those areas. The presence of

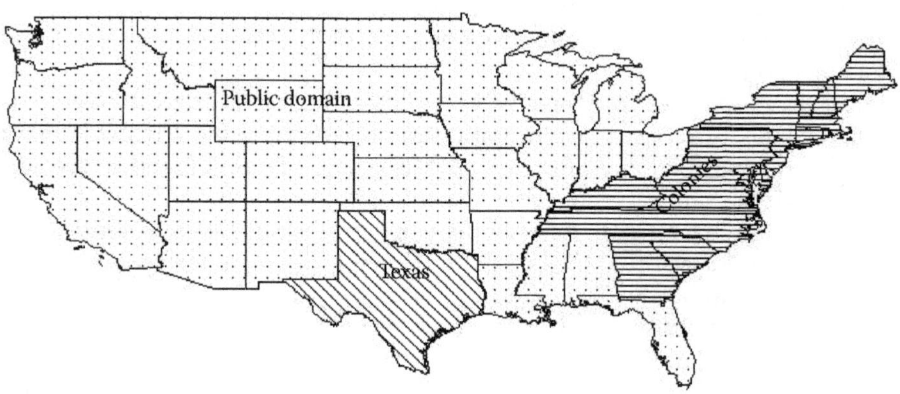

FIGURE 3.8
Public domain lands of the United States (Alaska not shown).

settlers would solidify the nation's claim to the area and help defend the area. In addition, revenue from the sale of the land would be a source of revenue for the new nation. Therefore, it was essential to conduct surveys to subdivide the land into sellable parcels and to provide for the orderly sale and settlement of the land. These public domain lands were typically surveyed prior to settlement using the rectangular land survey process known as the U.S. Public Land Survey System. That system subdivided the land into six-mile square townships with boundaries in the cardinal directions and further subdivided each township into 36 one-mile square sections.

For a time after the sale of public domain lands began with the Land Ordinance of 1785, land was sold in town-sized parcels to imitate the New England system of land development. This was in accordance with the argument of Treasury Secretary Alexander Hamilton with the goal of obtaining maximum revenue from the land. After it was realized that only speculators had funds for purchasing such large tracts, the method of selling public domain land gradually shifted to favor direct purchase by the settlers. At first, the minimum size for land sales was a section of land (640 acres). In 1800, to further assist settlers, the Congress dropped the minimum sale size to a half section (320 acres) and then to a quarter section (160 acres) in 1804. Ultimately, the minimum was reduced to a quarter-quarter section (40 acres), the amount of land considered necessary for sustaining a family, which inspired the slogan of "40 acres and a mule."

As each sale was made, a formal patent was issued by the local government land office and entries made in registers for the territory. Such patents serve as the basis of title to most lands within the states now composing the public domain lands.

It should be noted that certain lands in the public domain states were not conveyed in this process. Sovereign grants to individuals preceding acquisition by the United States were typically confirmed and accepted by the United States under terms of the acquisition treaty and never became part of the public domain. Navigable waters were reserved from the lands that were sold as part of the public land survey as highways of commerce and became the property of the public of each state with statehood. After passage of the Swamp and Overflow Land Act of 1850, lands that were deemed unfit for cultivation were also segregated by the surveys from the lands to be sold. Such swamp and overflowed lands were patented to the individual states, as opposed to private citizens. The states were, in this manner, provided the opportunity to reclaim the lands and then sell them to private individuals for revenue for the operation of state government. Additional lands were also granted to the states for public purposes, including Section 16 of each township, which was reserved for funding public schools. Grants were also made for the purpose of railroads and other functions to facilitate development of the territories.

Recommended Additional Reading on Land Tenure in the United States

Clawson, Marion (1964). *Man and Land in the United States.* Lincoln: University of Nebraska Press.

Clawson, Marion (1968). *The Land System of the United States.* Lincoln: University of Nebraska Press.

Linklater, Andro (2002). *Measuring America.* New York: Penguin Group.

McNeil, Kent (2000). "Aboriginal Title as a Constitutionally Protected Property Right" in *Beyond the Nass Valley: National Implications of the Supreme Court's Delgamuukw Decision.* Vancouver, BC: The Fraser Institute, 55–75.

Nelson, William E. (2008). *The Common Law in Colonial America.* Vol. 1. The Chesapeake and New England, 1607–1660. Oxford: Oxford University Press, Inc.

Platt, Rutherford H. (2004). *Land and Society.* Washington, DC: Island Press.

Pound, Roscoe (1938). "The Influence of the Civil Law in America." *Louisiana Law Review,* Vol. 1, No. 1, 1–16.

Powelson, John P. (1988). *The Story of Land.* Cambridge, MA: The Lincoln Institute of Land Policy.

Slattery, Brian (1992). "The Legal Basis of Aboriginal Title" in *Aboriginal Title in British Columbia,* F. Cassidy, ed. Montreal: Institute for Research on Public Policy, 113–129.

Watson, Blake A. (1982). "State Acquisition of Interests in Indian Land: An Overview." *American Indian Law Review,* Vol. 10, 219–256.

Watson, Blake A. (1998). "The Thrust and Parry of Federal Indian Law." *University of Dayton Law Review,* Vol. 23, 438–514.

Watson, Blake A. (2006). "John Marshall and Indian Land Rights: A Historical Rejoinder to the Claim of 'Universal Recognition' of the Doctrine of Discovery." *Seton Hall Law Review,* Vol. 36, 481–549.

Watson, Blake A. (2011a). "The Doctrine of Discovery and the Elusive Definition of Indian Title." *Lewis & Clark Law Review,* Vol. 15, 995–1024.

Watson, Blake A. (2011b). "The Impact of the American Doctrine of Discovery on Native Land Rights in Australia, Canada, and New Zealand." *Seattle University Law Review,* Vol. 34, 507–551.

4

Subdivision of Land in the United States

In the United States, the origins of most currently recognized title to land in the areas covered by the original colonies are royal grants from the sovereign of the several European nations claiming land based on "discovery." Those large royal grants were typically subdivided under various processes into smaller land tracts for use by the original settlers. For land acquired by the United States after the formation of the union, most of that land was subdivided under the regular process of the U.S. Public Land Survey System (PLSS) for use by settlers. In both cases, the original tracts have typically been subdivided into smaller tracts over time. As a result, a great variety of differing methods have been used for subdividing the country into currently recognized parcels.

4.1 Subdivision of the Original Grants to the United States

In New England, land was typically allocated to communities or church congregations and subdivided by a surveyor prior to settlement. As a result, settlement patterns were relatively regular, with no gaps left between lots within each township or the townships themselves (Figure 4.1).

In other colonies, differing approaches were used. For example, in Virginia, settlers were allowed to choose their parcel of land and then have it surveyed and registered. Thus, highly irregular settlement patterns resulted with gaps left between the more desirable tracts. In the Carolinas and Georgia, the proprietors of the original sovereign grants planned to presurvey their colonies into rectangles with lines run in cardinal directions. Despite such plans, the urgent demand for land resulted in a complete breakdown of the planned process. As a result, settlers defined their claims with blazed trees or raised mounds and described their holdings in terms of those landmarks for registration purposes, also resulting in irregular patterns of land ownership. See Section 4.5 for a detailed discussion of the varying methods by which land was subdivided in the colonial states.

FIGURE 4.1
Typical New England village plat. A 1749 six-mile square village plat in the English Provence of New Hampshire (now Bennington, Vermont) laid out by Surveyor Mathew Clesson (Truesdell, W.A., *The Rectangular System of Surveying*, 1908).

4.2 Informal (Private) Subdivisions

Many of the tracts of land settled by the original settlers of the United States have been informally subdivided over time. Such informal subdivision of land may occur with the division of land among the heirs to the property or when the owner decides to sell a potion or all of his or her land to others. When such informal subdivision occurs in the public domain states, the subdivided tracts may take the form of aliquot (fractional) portions of section or may be irregular areas within a section that is delimited by bearing and distance measurements referenced to one of the corners of the section. When informal subdivision occurs in other jurisdictions, the new tracts may be delimited by bearing and distance measurements referenced to other prominent landmarks or by designating adjoining parcels or features. For such informal subdivisions, plats are typically prepared solely for the parties involved in the transaction. Therefore, the plats are typically not made a part of the public records. The only record of the boundaries of the newly formed land parcels in the public records is typically the recorded deeds of sale.

4.3 Formal (Recorded) Subdivisions

In most jurisdictions within the United States, subdivision of land is now a regulated process. Under those processes, subdivisions must meet certain standards, and the resulting plat, after approval by the public officials, must be recorded and become part of the official public records (Figure 4.2). Typical restrictions for a subdivision include the parcel sizes being in accord with any official community planning requirements for the location; meeting various environmental requirements; addressing impact on transportation facilities, schools, and public safety facilities; and meeting prescribed criteria for roads within the subdivision. Often, the formal subdivision process involves a dedication of any new roadways to the local municipality. In such cases, the subdivision plat must be annotated to that effect and executed by both the conveying and the receiving parties. The plat then becomes a deed to the municipality for the road rights of way within the subdivision.

4.4 U.S. PLSS

In lands acquired by the postcolonial United States, land was typically surveyed prior to settlement using the rectangular land survey process known as the U.S. PLSS. That system subdivided the land into six miles square townships with boundaries in the cardinal directions and further subdivided each township into 36 one mile square sections. Since this survey created the basis for the vast majority of the United States, the development of this system and the subdivision process will be described in detail in the following sections. The history of this large division is a fascinating one, involving considerable trial and error, and with less than perfect results. Nevertheless, mathematical and legal accommodations have been successful in making and keeping it operational in a practical sense. But the history is incomplete without background knowledge of the system itself and what led to its design.

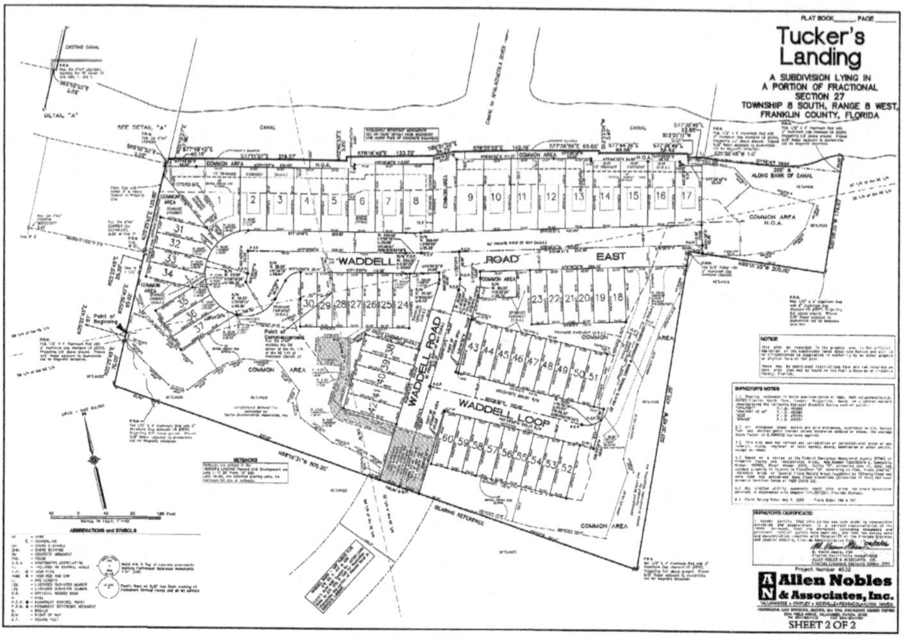

FIGURE 4.2
Typical recorded subdivision plat. (Courtesy of Nobles Consulting Group.)

4.4.1 Development of the U.S. PLSS

Several events led to the development of the U.S. Public Land Survey (White 1982): Following the Declaration of Independence in 1776, the 13 new states declared all English Crown lands and any unpatented lands of the proprietors of the colonies as the property of the various states. This included all of the unsettled lands to the west. In addition, the states confiscated lands of people who remained loyal to the English Crown. This resulted in the states becoming owners of millions of acres of undeveloped land. In addition, seven of the former colonies (Connecticut, Georgia, Massachusetts, New York, North Carolina, South Carolina, and Georgia) also had claims to unexplored and unsettled lands to the west based on their original charters or treaties with Native Americans. In addition to the fact that a number of those claims overlapped, the existence of those western lands presented a problem to those colonies without such lands. To solve those issues and for revenue for the new federal government, the Continental Congress requested that the new states relinquish their claims to those lands. (The Congress had no taxation power, was deeply in debt to France and other creditors from the war for independence and expenses of setting up the new government, and had no other assets with which to pay off the debts and operate the government, so the government sorely needed a source of revenue.) With some reluctance, the states did so, beginning with New York in 1780. Meanwhile, the United States negotiated a four-way treaty with England, Spain, and France in 1783 that resulted in ownership by the United States over all of the territory east of the Mississippi River, south of the Great Lakes, and north of Spanish Florida. This cleared the way for the sale of the lands ceded by the states to generate revenue. In addition, this created considerable controversy regarding the best method for selling the large amount of public lands.

One side of the controversy was composed primarily of small farmers, frontiersmen, and small businessmen from New England and the northern states. That group, represented by Thomas Jefferson and John Adams, advocated direct sale of land primarily to individual settlers in small parcels. To facilitate that goal, they supported comprehensive surveys of all the land before sale and the use of a rectangular system to prevent overlapping claims, gores, or gaps. They argued that such a system also provided for disposal of any undesirable land along with the more desirable land.

The second group, represented by Alexander Hamilton and John Jay, was generally composed of southern plantation owners and aristocrats. That group was more in favor of large grants to companies who would handle the surveying, sale of the land to individuals, and other business aspects of land distribution. They believed that systematic surveys prior to sale of the land would slow down sales. Rather, they advocated indiscriminate location with metes and bounds descriptions.

The Jefferson/Adams position prevailed, and the first design of an orderly system of a magnitude of dividing half the country was begun by Jefferson, a man who had surveying and mapping background and training in the law of the land and who was familiar with land patterns in Europe as well as having a working knowledge of the metric system of measurement. The day that Virginia's cession of its western lands was accepted, Jefferson, leading a committee of three, submitted to Congress an ordinance for the government in the western territories. The plan, approved by Congress, was known as the ordinance of 1784, which included the demarcation of state boundaries. These were to follow parallels of latitude, be of a specified size and rectilinear in form. Considering the newly contemplated states, the ordinance of 1784 also proposed dividing each state into hundreds with squares, each square being further divided into lots of one-mile square, with division to be accomplished with lines running north–south and others running at right angles to them. Due to this plan interfering with several states already having established their borders, this plan was later abandoned, and an action by James Monroe resulted in Congress repealing that part of Jefferson's plan.

More importantly, the ordinance authorized the survey and sale of a large territory now lying in the State of Ohio. The area to be surveyed was land that had been purchased from the Native American inhabitants. The ordinance required that the land be subdivided into seven ranges of townships of six miles by six miles, with lines run in cardinal directions, each subdivided into 36 lots containing 640 acres each. The ordinance directed that the lots were to be sold at an auction in each state at a cost of at least $1 per acre plus administrative costs of $1 per lot and surveying costs. The ordinance further specified that Lot 16, near the center, of each township be reserved for a public school, that four additional lots be reserved (possibly for governmental or religious purposes), and that rights be reserved to one-third of any gold, silver, lead, and copper found on the land.

The survey was begun in September of 1785 and the field work for the 84 complete or partial townships in the project area was completed in June 1787. As might be expected with a new process, less than desirable results were achieved. The Land Ordinance of 1785 did not address how field notes or plats were to be prepared, how corners were to be monumented, or the process for the subdivision of the townships. As a result, only the township corners and the lot corners along the township boundaries were monumented. Those corners were marked with wood posts or corner trees witnessed with two scribed bearing trees. The system of numbering of the townships used in this initial survey was confusing, and boundary lines were apparently not corrected for magnetic variation and some of the distance measurements were relatively crude. Nevertheless, the so-called "Seven Ranges" survey served as a learning experience and demonstrated that the rectangular system of townships and ranges was far superior to the indiscriminate location method in disposing of large tracts of land due to its obvious advantages including the simplicity of the resulting land descriptions.

The Ohio Company, a private land speculation company, petitioned Congress in 1787 to allow them to purchase a large unsurveyed tract of public land lying immediately west of the Seven Ranges survey. After debate, Congress agreed to sell approximately 964,000 acres to the company at about 12 cents per acre and donate another 100,000 acres to encourage settlement. The sale provided that the land would be subdivided into townships and lots per the Land Ordinance, with the company responsible for the survey, and that similar reservations of one lot for public school purposes, three lots for government use, and one lot for religious purposes would be made. The land would be sold by the company at a price of 67 cents per acre with credit available at 6% interest.

The survey was found to have similar problems as the Seven Ranges survey. Bearings of lines differed from cardinal as much as four degrees, distances were off as much as five chains in a mile, and crude plats were made but no field notes are known to exist for the survey of the Ohio Company tract.

In 1788, a year after the sale to the Ohio Company, Congress agreed to sell 1,000,000 acres in the same general region and with the same terms to John Cleves Symmes, a major contributor to the Continental Army during the Revolution. As with the Ohio Company survey, the survey for the Symmes tract was badly managed and effectively stopped future sales of large land tracts. It reinforced the need for properly supervised and tightly controlled surveys in the disposal of the public land.

In addition to the problems with the early surveys, the revenue generated by sales of public land proved to be disappointing. Settlers were not willing to pay a dollar per acre cash for land when they could buy it for considerably less, on credit, from either the Ohio Company or Symmes. The main source of revenue from the land had been the two large tracts that were sold at a few cents per acre. As a result, Congress was anxious to proceed with a more sound approach. The Secretary of the Treasury, Alexander Hamilton, was asked to develop a plan for this. Hamilton did so with a plan for selling the land at 30 cents per acre with two years of credit at 6% interest offered. However, Congress did not act on his plan, or any plan for surveying and selling the public land, until several years later, when the 1796 Land Act was passed. That act directed the survey and sale of all of the land northwest of the Ohio River and above the Kentucky River in Kentucky, where Native American title had been extinguished by treaty, including all of the unpatented land in the Seven Ranges, Ohio Company, and Symmes tracts. The new Land Act provided standards that would eliminate many of the problems experienced with the previous public land surveys. This included authorization of the appointment of a surveyor-general, with clear authority over the surveys; required that the six-mile square townships have north–south lines run on true meridian and for east–west lines to cross them at right angles; required monumentation of all township and section corners; required sections to be numbered starting in the northeast corner and then running alternately east and west; required detailed field notes and plats; specifically required that a Gunter's chain be used; and designated all navigable rivers in the territory as reserved as public highways. The new act also increased the price of the land to $2 per acre and provided for credit purchases.

Although the 1796 act essentially established standards reflected in most of the later Public Land Surveys, there have been various changes in the standards and procedures from time to time with new direction from Congress or from instructions issued periodically by the surveyor-general. Also, as the United States acquired new territories, the survey was authorized to be extended to those territories. As a result, the U.S. Public Land Survey now has been extended to cover the majority of the area of the country and serves as the root of title for most of the land in all of that area.

4.4.2 The Subdivision Process

The public land subdivision process involved the following steps for most of the surveys:

1. In each territory, an initial point was selected as a point of origin for the survey system and two axes run from that point (Figure 4.3). The north/south axis, called the principal meridian, was run true (astronomic) north and south. The east/west axis was called the baseline.

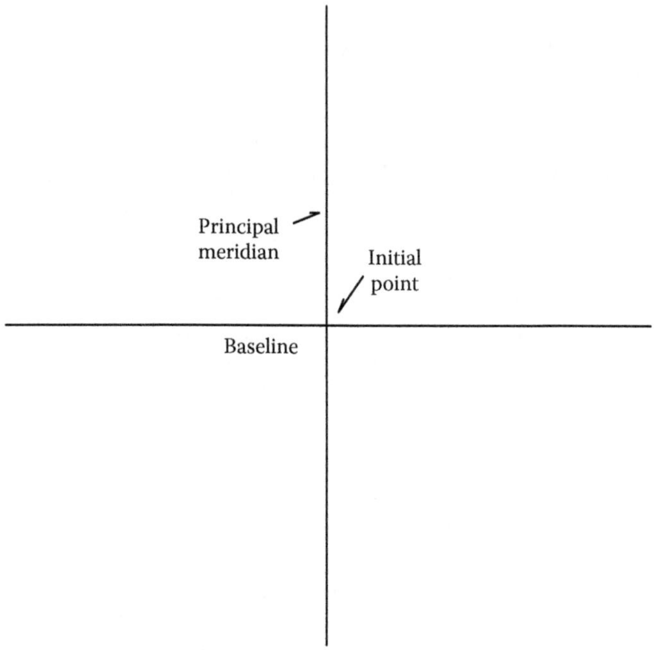

FIGURE 4.3
Baseline and principal meridian.

2. Secondary axes, called standard parallels, were established at intervals of 24 miles north and south of the baseline. In addition, secondary axes, called guide meridians, were established at intervals of 24 miles east and west of the standard parallel. All such standard lines were run at astronomic cardinal directions. These standard lines forming a grid at 24-mile spacing are also called correction lines since the primary purpose of the lines is to allow correction for convergence of the meridians. Note, in Figure 4.4, that since the guide meridians are run at astronomic north, they tend to converge toward the principal meridian within each 24-mile grid. At each standard parallel, a new guide meridian was begun at an offset of 24 miles. Thus, there were double corners set on each standard parallel at the intersections with the guide meridians, one corner pertaining to the guide meridian to the north and one pertaining to the guide meridian to the south.
3. Each of the 24-mile grids formed by the standard lines was subdivided into six-mile square townships (Figure 4.5). This was accomplished by running range lines in an astronomic north or south direction at six-mile intervals along each standard parallel. Each range line extended from one standard parallel to the next. Township corners established at six-mile intervals along each range line were then connected by township lines run in an east–west direction.

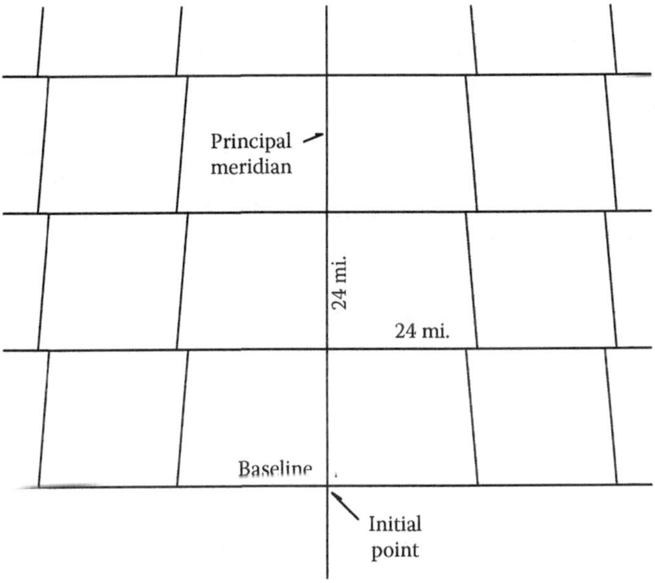

FIGURE 4.4
Standard lines. Note: Convergence of the meridians is exaggerated.

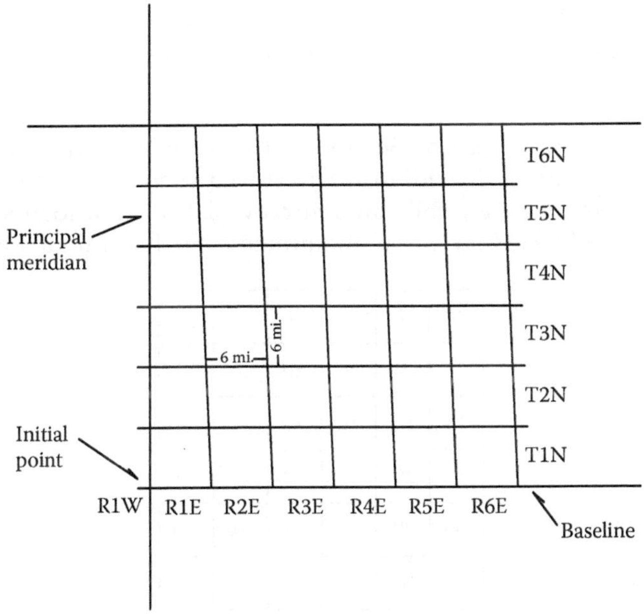

FIGURE 4.5
Subdivision of standard grids into six-mile square townships.

For purposes of identification, each horizontal tier of townships is identified by the position of the tier northerly or southerly of the baseline. Similarly, each vertical column of townships, or range, is identified by the position of that range east or west of the principal meridian. Thus, each township has a unique identifier similar to a Cartesian coordinate. For example, a township lying in the second tier southerly of the baseline and in the third range easterly of the principal meridian would be identified as Township 2 South, Range 3 East.

4. The final step in the standard survey process was the subdivision of each township into one-mile square sections. This was accomplished by running, at intervals of one mile, meridian lines parallel to the east boundary of the township. Then, section corners set at one-mile intervals along each such meridian line were connected by lines running easterly or westerly. The result is that each standard township is divided into 36 one-mile square sections.

   The sections were numbered consecutively running east to west and west to east, beginning with Section 1 in the northeasterly corner of the township (Figure 4.6). Since the standard section is one mile square, it contains 640 acres. Note that the meridian lines are not necessarily cardinal but are run parallel to the east line of the township. As a result, the western tier of sections may be slightly smaller than a standard section due to convergence of the meridians. The general policy was to place all such deficiency (or any excess) in the western and the northern tier of sections in each township.

Thus, a description such as Section 6, Township 1 North, Range 2 East, Tallahassee Baseline & Meridian referred to a section of land that can be readily identified on the public land survey plat for that township as well as on the ground based on the set monuments. Such a unique identifier can

6	5	4	3	2	1
7	8	9	10	11	12
18	17	16	15	14	13
19	20	21	22	23	24
30	29	28	27	26	25
31	32	33	34	35	36

FIGURE 4.6
Numbering system for sections in a typical township.

be used for conveyance of the land and other purposes in the legal system. Therefore, it allowed for the orderly conveyance and management of public land of the United States.

The previously described process is the standard for the public land surveys. At times, it was necessary to deviate from that standard processes when excepted lands, such as grants preceding the subdivision, large water bodies, or other obstructions, were encountered.

When dealing with navigable waters, the public land survey process called for separating the upland, subject to public sale, from the navigable waters, which were to be preserved as public highways. The separation process involved meandering, or running a series of measured courses and distances around the perimeter of the water body. The alternate course of action, for nonnavigable water bodies, was to continue section lines across the water bodies. For such nonnavigable water bodies, the width was usually noted in the field notes and the approximate limits of the water sketched on the plat, as opposed to an actual survey or meander of the shoreline, as required for navigable water bodies.

Sections of land bordering navigable waters were often divided into smaller lots or fractional portions of sections and numbered as government lots. Such smaller parcels, less that one section in size, were not surveyed in the field, but their graphic depictions and calculated area noted on the plat did allow for sale of the more valued riparian lands in smaller parcels.

4.4.3 Measurements Standards

During the debate to establish standards for the U.S. Public Land Survey, Thomas Jefferson advocated the use of a geographic mile as the standard. That distance is the length of one minute of arc on the equator, or roughly 6080 feet. That unit would have resulted in sections having an area of about 849 acres, an irregular figure. Therefore, a chain with a measurement resulting in exactly 80 chains per statute mile was adopted to provide an area for sections of 640 acres.

Therefore, the basic unit of measure adopted was the Gunter's chain. That instrument, used for distance measuring during the early phases of the survey, was a linked chain having a hundred links with a total length of 66 feet. Thus, each link represented 0.66 feet or 7.92 inches. The length of the chain was convenient in that 80 chains represented an even mile, the length of a side of a section.

4.4.4 Monumentation Standards

The monumentation standard used in the public land survey was to set a monument at each section corner and at each quarter corner. Thus, the perimeter of all sections had monuments at one-half-mile intervals. Along meander lines, monuments were set at each intersection of the meander line and section lines. In addition to the monuments themselves, distances and bearings were taken to witness trees at each monument to perpetuate the location in the event of the loss of the monument. Witness trees were prominent trees nearby the monument that were blazed and scribed with the identification of the corner. Four witness trees were established for each section corner, while two were established for each quarter corner or meander corner. The monuments themselves varied. In the early portions of the public land survey, scribed wooden posts were used in forested areas. In other areas, mounds of earth or stone or pits were used. In more recent times, concrete monuments with brass caps and iron posts with caps were used.

4.4.5 Field Notes and Plats

Field notes for the public land surveys represent a fairly complete record of what the original surveyor measured and saw on the ground as the lines were run and were intended to provide a complete description of each section to facilitate the conveyance of the public lands. Earlier field notes (Figure 4.7) were typically in script that often presents a challenge in interpretation (Figure 4.8).

The intent of the notes was to provide not only information regarding the survey measurements but also information regarding the nature and value of the land being surveyed. Although some public land surveyors kept more detailed notes than others did, the following is a list of the more commonly required information that was either required or requested to be noted, over much of the 19th century (Johnson and Smith 1913).

- The precise length of every line run
- The variety and diameter of all bearing trees, with the direction and distance of the same from their respective corners
- The kinds of materials of any monuments set
- Identification of trees on line

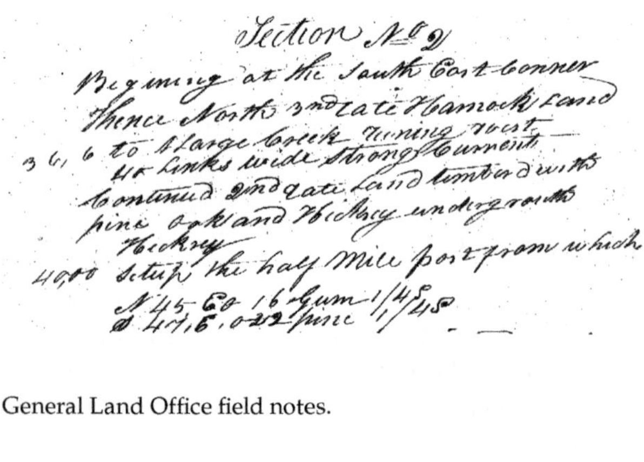
Section No 9
Beginning at the South East Corner
Thence North 3rd rate Hamock Land
36,6 to A Large Creek running West
40 Links wide Strong Current
Continued 2nd rate Land timbered with
pine Oak and Hickory undergrowth
Hickory
40,00 Setup the half Mile post from which
N 45 E 0 16 Gum 1/45
S 47 E 0 22 pine 1/48

FIGURE 4.7
Typical early General Land Office field notes.

East of Section No. 31
Began at the SE corner of said section from
Thence N
10.00 Dry land
Land 2 rate Timber Black Oak Pine woods
Open
40.00 1/2 mile post from which stands
40.15 quarter section post
S56W 41 Hickory/S 1/4
S55E 38 Black Oak/S 1/4
Land and timber the same as before
76.00 Pond
80.26 To the NE Corner of said section from

which stands

N 22 1/2 E 5.00 Pine/S29 T1 R2
N 6 1/2 W 4.30 Live Oak/S30 T1 R2
S 79 W 1.40 Pine/S31 T1 R2

FIGURE 4.8
Interpretation of field notes in Figure 4.7.

- Intersections of line with land objects such as prairies, streams, swamps, and marshes with the course of the same at the points of intersection
- In cases of navigable streams, their width will be ascertained between the meander corners
- The land's surface—whether level, rolling, broken, or hilly
- The soil—whether first, second, or third rate
- Types of timber and whether first, second, or third rate
- Springs of water—whether fresh, saline, or mineral, and the course of the stream flowing from them
- Lakes and ponds—describing their banks and giving their height, and also the depth of water, and whether it was pure or stagnant
- Improvements such as towns and villages, Indian towns, houses or cabins, fields, or other improvements
- Evidence of minerals such as coal, peat, and ores, also salt springs and licks
- Roads and trails, with their directions
- Rapids, cataracts, cascades, or falls of water, with the height of their fall
- Precipices, caves, sink-holes, ravines, stone quarries, ledges of rocks, with the kind of stone
- Natural curiosities, interesting fossils, petrifactions, organic remains, also all ancient works of art, such as mounds, fortifications, embankments, ditches, or objects of like nature

A plat was prepared for each township in the public land surveys (Figure 4.9). Typically, these were plotted at a scale of two inches to the mile. The plats usually depicted the bearing and distances of all of the section boundaries as well as various topographic features such as water bodies, wetlands, and existing cultural features. When navigable waters were in the township, the plats typically contained a table of bearing and distances for all of the courses of the meander line.

4.5 Other Rectangular Cadastral Surveys in the United States

Rectangular systems have long preceded the design and implementation of the U.S. PLSS. As the beginnings of American law relied heavily on European decisions and concepts, so did the layout and development of the wilderness of the New World encountered by Europeans. The Roman Empire, once including much of western Europe, developed a rectangular system of land and parcel organization known as the Roman *centuriatio*. This has been characterized as "perhaps the oldest surviving form of land use survey in the Western World." Present-day Tunisia, established after the fall of Carthage in 140 BC, began a period of complete land survey and assignation of individual land parcels. Similar surveys have been found in parts of Italy, the Po lowlands, and in southern France.

Studying this part of history, one will find a system of land utilization that is simple, straightforward, and rational and works efficiently enough for it to survive, unchanged, for over 20 centuries. One will also find the layout of fields, roads, drainage ditches, and nearly all other works of man on the landscape accommodated in the rectangular grid. Roman maps demonstrate widespread use of the pattern throughout much of western Europe and elsewhere where the Roman Empire had exerted its influence.

Rectangular grants, lots, and parcels were often the preferred configuration. They were simple to create and lay out and even simpler to divide. The only obstacle to complete rectilinearity of parcels is topography, where rivers, lakes, and mountain crests have become the preferred natural boundary. Even so, where one boundary of a grant was a nonstraight component, usually the remaining boundaries (often three to complete roughly a square or rectangle) were straight, sometimes influenced in orientation by the natural boundary.

Perhaps more fascinating than the history of the PLSS is the history of various other rectangular systems in the United States, much of which occurred prior to the Ordinance of 1784. It involves the division of all 13 colonies, Texas and Hawaii, along with scattered grants and homesteads in many of the other states. Each of the states, or territories, implemented one or more types of rectangular systems, depending on design for settlement, topographical features, or to allow for land grants made prior to the formal division.

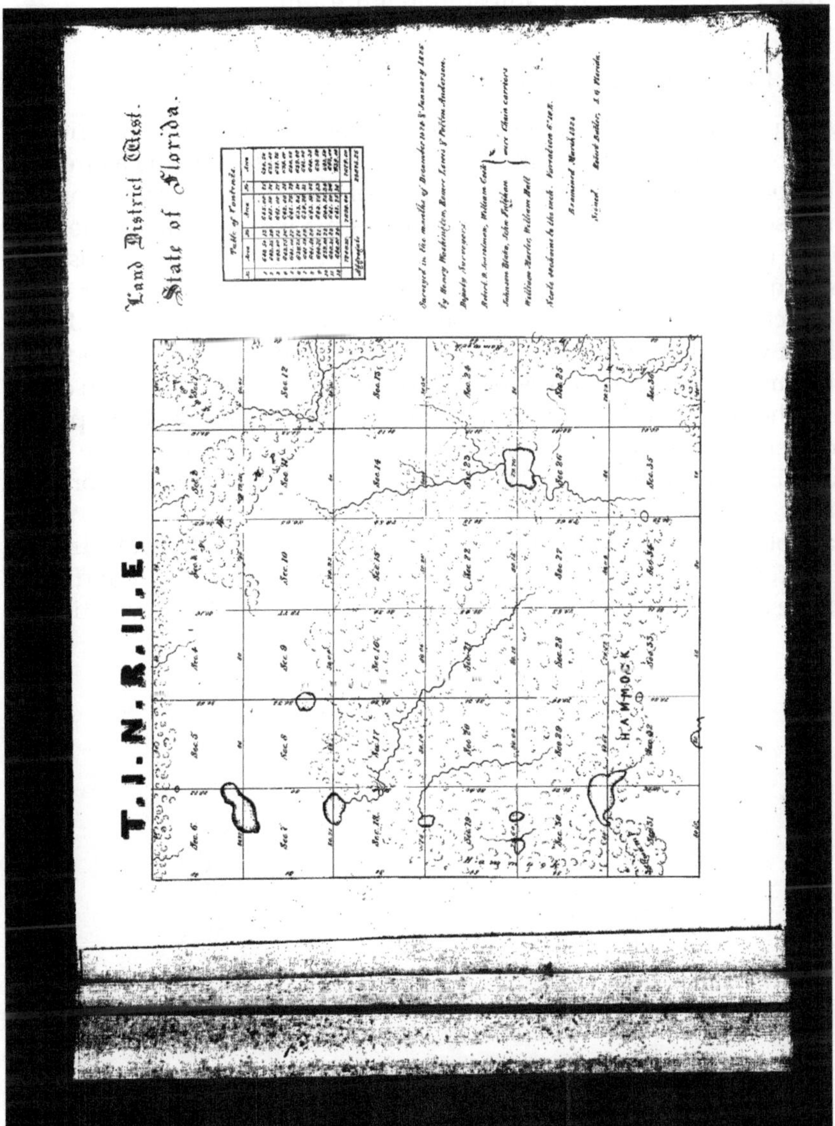

FIGURE 4.9
Typical public land survey plat.

According to Truesdell (1908), Bennington, Vermont (chartered in 1749), was the first township in America to be laid out six miles square with boundaries north, south, east, and west. It was a grant made by the Governor of the Province of New Hampshire, Benning Wentworth, in what is now Vermont, at that time claimed by New Hampshire (Figure 4.10).

4.5.1 The 13 Original Colonies

To have a full understanding and appreciation of the beginnings of land tenure with its accompanying granting of lands and subsequent measurement and division, it is important to have a grasp of government affairs from the time of discovery. Different European powers made their claims, the most significant of which were the British, the French, the Dutch, the Swedish, the Spanish, and the Russian governments. Each had its own methods of settlement, which eventually gave rise to several different systems of land divisions, each with different measurement systems to define their boundaries. Some had more than one definition of measurements within their own development scheme, the *rod*, the *arpent*, and the *vara* being common examples.

There were three types of colonies and systems of government in the North American British Colonies: Royal Colonies, Charter Colonies, and Proprietary Colonies.

Royal Colonies were ruled directly by the English monarchy and administered by a royal governor, a council appointed by the British crown, and a representative assembly elected by the people. At the time of the Revolutionary

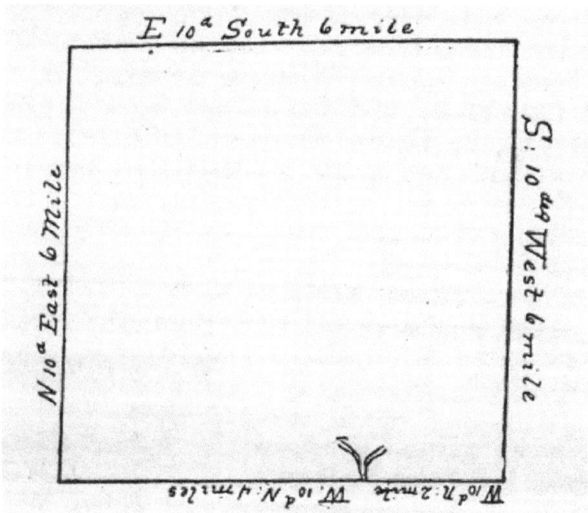

FIGURE 4.10
Survey plat of the exterior boundaries of first township in America to be laid out six miles square with boundaries north, south, east and west (Truesdell, W.A., *The Rectangular System of Surveying*, 1908).

War, the Royal Colonies were New Hampshire, New York, New Jersey, Virginia, North Carolina, South Carolina, and Georgia.

Charter Colonies were generally self-governing, and their charters granted to the colonists as opposed to proprietors. They were controlled by written contracts between the reigning king and the American colonists, defining the share that each should have in the government, which could not be changed without the consent of both parties. A charter bestowed certain rights, privileges, or franchises from the sovereign power to the colony. The Charter Colonies were all located in New England. At the onset of the Revolutionary War, the Charter Colonies were Massachusetts (Charter then later Royal), Connecticut, and Rhode Island. The first Charter in North America was granted to Sir Walter Raleigh by Queen Elizabeth I. Raleigh sent his first colonization mission to the island of Roanoke, which is in present-day North Carolina.

Proprietary Colonies were territories granted by the English Crown to one or more proprietors, similar to a corporation and often made up of speculators, many of whom never left Europe, who had full governing rights. These were the types of settlements in British Colonial America established between 1660 and 1690. King Charles II established the first Proprietary Colonies in order to repay debts and favors. The British Crown awarded their most loyal supporters with large tracts of land in colonial New York, New Jersey, Pennsylvania, Maryland, Delaware, and the Carolinas. These proprietors were given authority to supervise and develop the colonies into successful and profitable enterprises. The lands were quickly colonized with British subjects at the expense of the proprietors. By the time of the Revolutionary War, there were three Proprietary Colonies: Pennsylvania, Maryland, and Delaware.

These three types of colonies and systems of government in the North American British Colonies would sometimes change in status depending on political and economic changes in Great Britain, or with changes in controlling European power. The government of a self-governing Charter Colony or a private proprietary Colony therefore could become a Royal Colony when the power reverted to the king. The colonies of New York, New Jersey, North Carolina, and South Carolina began as proprietary colonies but later became royal colonies.

4.5.1.1 Virginia

Virginia was the first colony established in Colonial America. King James I claimed all the land in America between the French settlement on the St. Lawrence and the Spanish settlement in Florida. He issued the first charter in 1606 to Sir Thomas Gates and three others, known as the London Company. Included was a similar grant to Thomas Hanham and three others, to be called the Plymouth Company. The Virginia Colony was founded by Captain John Smith with other colonists by the London Company, at Jamestown, in 1607.

The Virginia Company of London granted land patents to settlers until 1623. In that year, the Company was dissolved and the Crown then distributed land through the office of the secretary of the colony. There were two types of grants—headright, which was established in 1627, and straight purchases. After the Virginia Company was dissolved, settlers could purchase patents through a clerk of the county until 1921.

In 1649, King Charles II granted the Northern Neck area between the Potomac and Rappahannock Rivers to seven of his supporters, one of which was Thomas, Lord Culpeper. Over time, Culpeper purchased the shares of the others and by 1690 became associated with Thomas, Lord Fairfax, and began granting from the more than 5 million acre proprietorship.

Virginia became a royal colony in 1624. By 1654, millions of acres were granted to individuals claiming headrights during the 17th century. The Virginia land office opened in 1779 and began approving languishing land claims and processing military warrants. At this time, West Virginia was part of Virginia along with what later became Kentucky.

After the Revolution, the federal government asked Virginia and other states to cede their western lands. In 1781, Virginia relinquished its claim to lands in the Northwest Territories in exchange for being able to award bounty lands in the Virginia Military District in what is now south-central Ohio, which it did until Ohio became a state in 1803. Land in 23 Ohio counties from the Ohio River north between the Scioto and Little Miami rivers was reserved by Virginia to satisfy its military bounty warrants. That area, known as the Virginia Military District, is one of the original nine major subdivisions in Ohio and the only one not using a rectangular survey system. In 1852, Virginia ceded all unclaimed lands to the federal government, which in turn ceded these remaining lands to Ohio in 1875.

The area now within the state of West Virginia was part of Virginia until 1863, when it became a state. Many grants were made by Virginia, followed by West Virginia after it became a separate state. In 1774, Virginia offered 1000 acres for each family brought in as encouragement to settle in Western Virginia. Many of the speculators formed land companies, who surveyed and sold to individuals, who then obtained title by patent. By 1754, over 2.5 million acres had been granted to these companies. The first warrants for military bounty land in this area were issued in 1782. Many soldiers sold their warrants to speculators, who in turn sold to others.

The area now within the state of Kentucky was also originally part of Virginia until it became a separate state in 1792. Prior to that date, Kentucky proprietors obtained their patents from Virginia, and their rights were of three kinds: military bounties, preemption and settlement, and warrants from the treasury. Although considered a metes and bounds type of grant system, many of the original lots were rectilinear, at least in part.

The Kentucky area of Virginia includes the Jackson Purchase, a region bounded by the Mississippi River to the west, the Ohio River to the north, and the Tennessee River to the east. Although technically part of Kentucky at the time of statehood in 1792, the land did not come under definitive U.S. control until 1818, when Andrew Jackson purchased it from the Chickasaw Indians. Jackson's purchase also included all of Tennessee west of the Tennessee River. A number of Revolutionary soldiers occupied this region prior to the treaty. In 1820, the Kentucky General Assembly approved the surveying of military claims west of the Tennessee River. At a result, 242 patents were authorized and filed prior to 1792. Also in 1820, the Kentucky General Assembly passed legislation that declared the public land system would be used for mapping this region by Townships, Ranges, and Section. It was surveyed and mapped by William Henderson. Acreage was divided into square townships, each containing 36 square miles, and townships were then divided into 36 sections of 640 acres each. Each 640-acre section was then divided into quarter sections of 160 acres each. Beginning at the intersection of the principal meridian and the base of the Jackson Purchase and running East and West, the blocks were assigned range locations.

4.5.1.2 New York

New York was the second colony established in Colonial America. Founded by the Duke of York and other colonists on Manhattan Island, the Colony was classified as one of the Middle Colonies. There were three periods to the settlement of New York, each with its own systems of land division. The Dutch Period extended from 1624 to 1664.

The first colonial activity in what was known then as New Netherland was by the Dutch West India Company, a group more interested in trade than in colonization. Its 1629 charter provided for the neofeudal status of *patroons* for a stockholder who would establish a colony, a grant that fronted on the Hudson River and extended inland a specified distance. Patroonships extended from the mouth of the Hudson, upriver about 30 miles, with three additional claims further upriver. Due to a variety of circumstances, only two of these patroonships existed when the English took over New Netherland, Rensselaerswyck, which originally included the present site of Albany, and Colendonck, including Yonkers. The latter had few settlers and was subsequently divided into two English grants. Rensselaerswyck, granted in 1629, operated for two centuries and was the only one that had a significant number of occupants.

The Dutch also granted farms and townships, which accounted for immediate settlement of a greater number of people. Most of these farms were within what is now the limits of New York City and immediate surrounding areas. The township grants were mostly on Long Island. The island of Manhattan was initially reserved for the Company, which early on laid out about a dozen *bouweries* (farms) to company officers. The southern portion of Manhattan Island was divided into streets, blocks, and city lots. Towns, mostly settled by New Englanders, were divided according to English custom, mostly with longlots (see Section 4.3.5 for more discussion this topic).

The Dutch had its own definition of the rod for land measurement, different from the English rod(s). The English rod varied from 12 to 22 feet in length, depending on the quality of land being measured (the lower the land quality, the longer the rod used to measure it, resulting in deliberate overrun in total area). The Dutch rod is frequently taken to be 12 1/2 feet in length, although there was variation in the length of the Dutch foot. In addition, six Dutch miles is equivalent to 24 English miles.

The English Period extended from its takeover in 1664 until the Revolution, in 1776. Prior to 1685, land grants were few and were mostly regranting of earlier Dutch holdings. Grants of various sizes were then made by succeeding governors. Large grants, known as manors, were made along the Hudson River. Extensive patents were granted inland after the lands along the river were in private hands. Smaller grants were also made to farmers and as military bounties for service in the French and Indian War.

The biggest influence on the land system of New York came during the 18th century from Cadwallader Colden, who held two influential offices from 1720 to 1775, those of surveyor-general and, later, lieutenant governor. From 1720 to 1760, he was the colonial official concerned with the surveying and distribution of New York lands.

At the end of the Revolution, more than half of New York lands had not been granted. One large military tract for veterans, along with several private acquisitions, accounted for most of the western and northern parts of the state. Mostly, these lands were divided into townships in rectilinear fashion. Smaller tracts were granted in a variety of ways. Robert Morris acquired large tracts consisting of 1,000,000 or more acres. The subdivision of these large tracts was mostly geometric and commonly in rectangular grids. Many of these tracts were oriented with the cardinal directions. Some were surveyed into square townships and sections, and some with water frontage laid out in longlots.

Three large purchases in the western part of the state were the Morris Reserve, the Pulteney Purchase, and the Holland Land Purchase. Robert Morris, one of the signers of the Declaration of Independence and a financier of the American Revolution, purchased his tract from Massachusetts Bay and, subsequently clearing the title with the Seneca Indians, sold to investors.

In 1788, Oliver Phelps and Nathaniel Gorham purchased all of Massachusetts's preemptive right to land in Western New York, some 6,000,000 acres, called the "Phelps and Gorham Purchase." The United States forced the previous occupants, nations of the Iroquois Confederacy, to cede the lands as they had been allies of Great Britain. After the Massachusetts claim was settled, New York state intended to sell the land for development and settlement.

Phelps and Gorham were to pay $1,000,000 in three equal annual installments. Under the terms of the purchase agreement, they took title only when they had extinguished the Indian title. Later in 1788, they were able to extinguish Indian title to all lands east of the Genesee River between Lake Ontario and the Pennsylvania border, as well as a tract 12 miles by 24 miles paralleling the west bank of the Genesee River ("The Mill Yard Tract"), which amounted to about 2,250,000 acres.

In 1790, Phelps and Gorham became unable to pay the second installment on the purchase contract, and the preemptive right to lands west of the Genesee River reverted to Massachusetts. The state resold the right to Robert Morris while Phelps and Gorham received the deed to the lands east of the Genesee, to which they had extinguished title, but they conveyed all of the remaining unsold land to Morris.

The Pulteney Association was a small group of British investors who purchased a large portion (12,000,000 acres east of the Genesee River) of the Phelps and Gorham Purchase in 1792 from Morris's London agent, William Temple Franklin, who was a grandson of Benjamin Franklin. After Sir William Pulteney's death in 1805, the tract was known as the Pulteney Estate, which was managed by a series of agents. Some of their heirs owned land in western New York into the 1920s, with the last parcel of The Pulteney Association property, 10 acres, being sold in December 1926.

The Holland Purchase was made by a group of Dutch investors from Amsterdam. Since it was forbidden for foreigners to own land within the United States, the investors placed their funds in the hands of trustees who bought the land in central and western New York and western Pennsylvania. They purchased the 3,250,000 acres in 1793, divided and sold the lands, and dissolved their partnership in 1846.

Joseph Ellicott, a brother of Andrew Ellicott, assisted the latter with the survey of Washington, DC, and was hired by the Holland Land Company to survey and divide their purchase. He, with his brother Benjamin and 130 men, surveyed the purchase for the next three years. Fifteen ranges or tiers of townships, each six miles square, were laid out. These were divided into lots of 320 acres each. Where necessary, these lots were subdivided into 120-acre tracts.

William Bingham, the Philadelphia lawyer and banker, also invested heavily in New York lands. Binghamton is named after him as a purchaser of the original patent of 10,000 acres.

While Governor Wentworth of New Hampshire was granting lands in what is now known as the state of Vermont, New York also issued patents in the same area. However, in contrast to the New Hampshire grants, the New York patents were generally irregular in shape and issued to wealthy landowners. Between 1765 and 1776, the colony of New York issued 107 patents of town-sized parcels of land in what is now within Vermont. Most were issued after New Hampshire ceased selling its own grants of Vermont land. Twenty-four of them never received town names. Of the remaining 83, only 5 towns remained with their names unchanged. Twenty more were reissues of New Hampshire grants (see discussion regarding Brattleboro). The remainder 58 patents were tiny short-termed settlements. Since they only existed on paper, they were called "Paper Towns."

4.5.1.3 Massachusetts

Massachusetts, the third colony established in Colonial America, was founded by John Winthrop and other Puritans at Massachusetts Bay. In 1620, the Plymouth Colony was established, which included the area around Plymouth and Cape Cod. In 1630, the Massachusetts Bay Colony was founded, which expanded to north of Boston, including the eastern part of what is now New Hampshire. In 1652, Maine became part of Massachusetts, and in 1691, a new charter was granted to the Massachusetts Bay Colony, uniting it with the Plymouth Colony. Parts of Maine and Nova Scotia were also added to Massachusetts. In 1786, The Ohio Land Company was formed, which resulted in many Massachusetts residents emigrating to Ohio.

During the colonial period, land was granted by the General Court to proprietors, who also granted land. After the proprietors made their conveyances, or when a town was incorporated, town clerks recorded deeds, mortgages, and other title instruments. Like most other states, Massachusetts made grants in lands she either had jurisdiction over or otherwise claimed. As a result, a few of those grants were made in the area now within Vermont, and interestingly, most of Massachusetts land granting took place in that area of Massachusetts now known as Maine.

The first land patent in what is now the state of Maine was in 1622, when Sir Ferdinando Gorges and John Mason, president and secretary, respectively, of the Council of New England, granted to themselves jointly the land between the Merrimac and Sagadahoc (Kennebec) rivers. This had been a royal patent granted by King James I. In 1629, they divided the land at the Piscataqua River, Gorges retaining Maine, and Mason, the land south of the river as the Province of New Hampshire. In 1639, Gorges obtained a renewed patent for the area between the Piscataqua and Kennebec Rivers, in the form of a royal charter from King Charles I.

Beginning in the 1640s, the Massachusetts Bay Colony began claiming lands north of the Merrimac River, resulting in early settlements of the area that later became New Hampshire. In 1691, William and Mary issued a charter for the new Province of Massachusetts Bay that included, along with other lands, the former claims of the Massachusetts Bay Colony and the lands to the north previously granted in 1664 by Charles II to the Duke of York. The entire area remained part of Massachusetts until Maine became a state in 1820. In 1700, Massachusetts created the first Committee for the Sale of Eastern Lands, including Maine. In 1783, it created the second Committee for the Sale of Eastern Lands, specifically to survey, appraise, and sell land in Maine. In 1819, Maine separated from Massachusetts, but Massachusetts retained ownership to half of the available land, with Maine acquiring that remaining land situated within Maine borders from 1820 to 1853. In 1824, the Land Office of Maine was established to distribute this land and held auctions to dispose of it.

Because Maine was part of Massachusetts until 1820, it did not enjoy independent colonial status like its parent state did. However, many of the early grants were made by Massachusetts, and, except where influenced by topography, which consisted mostly of the many rivers, townships were mostly rectangular. In fact, the northern part of the state was laid out westerly of a baseline, which was the easterly boundary that became the U.S.–Canada border. These townships were mostly six miles square, many without names, but designated with T-R (township and range) identification. Much of this large area became either potato farmland or timberland, and today, most townships retain their original numbered designation.

An interesting facet of these township grants is that, in many, lots were reserved for church, minister, ministry, and sometimes school. Generally, these reserved lots fit the lotting pattern of the township and could be located anyplace within the township, but usually the location was designated on the lotting plan of the town. In other cases, reserved lands were noted in the proprietors' records and not always shown on the plan of the township. In addition to public, or common, areas, strips of land set aside for future roads were also made, again not always indicated on the plan itself.

Lotting plans were not consistent or according to a particular design, and the number of lots in a subdivision often depended on the number of grantees or the proprietors' wishes. There was no formal set of instructions, so there may have been a very few lots or 100 or more in a township. Numbering of the lots can also be a challenge, since a township may have been divided over a period of time, whereby it was broken down into divisions depending on land quality and intended use. Numbering was repeated from division to division, so there could be several lots in the town having the same number, but, fortunately, the plan and its subsequent conveyancing would state the name or number of the division in addition to the number of the lot, thus giving it a unique location. In those towns where no lotting plan is extant, the surveyor's challenge is often a complete reconstruction process, based on title documents and what field evidence can be uncovered.

The Massachusetts grants consisted of towns, or smaller areas, granted to entice settlement, such as land for a mill site, or a ferry landing. Relief to survivors of the several French and Indian Wars as well as grants to soldiers for military service were common. After statehood, Maine executed its own grants, a town at a time, which were later subdivided according to proprietor's wishes with the aid of their surveyor(s).

There was considerable speculation in Maine lands, mostly for timber, since there was a big demand in Europe for lumber, especially timber for ship masts. Around 1900, the city of Bangor was recognized as the lumber capital of the world. A few small groups of investors and some individuals acquired large areas of land for speculation. The foremost example is William Bingham, a Philadelphia lawyer who acquired two large tracts consisting of 1,000,000 acres each, which were subdivided into smaller units of rectangular units of towns except where topography required otherwise. In fact, Bingham's overall design contained not only square towns but also squares further subdivided into smaller squares, continuing down to lots of 10 acres in size, and containing provisions for future roads to provide access.

Many states have been subject to Native American land claims in recent past, and Maine is no exception. Their claim was settled in 1980 after four years of negotiations, with the tribes receiving $81.5 million, which was the largest settlement of its kind in the country and the first to include provisions for the reacquisition of land. Aboriginal rights are the basis for the trust relationship between the Wabanaki People and federal government. The Wabanaki People inherited their rights as a result of their families living here before Europeans laid claim to the land. Many of their rights were not given up in treaties and agreements and therefore still exist and are acknowledged by the United States and Canada (see Section 3.1.7 on Aboriginal Title).

The northerly boundary of Massachusetts was not settled until 1740. For many years, Massachusetts claimed, as interpreted by her first charter, three miles north of every part of the Merrimac River, which at times meant an extension as far north as the outlet of Lake Winnipesaukee in the middle of New Hampshire. Previous to 1740, Massachusetts had made many land grants in part or wholly north of the boundary set at that time. Several became towns, or eventually parts thereof, while several other grants were made to individuals.

Massachusetts made three township grants in what is now Vermont. The first grant was Northfield in 1672. A Native American deed was obtained in 1671 for a tract of land on both sides of the river, which was followed by additional Native American deeds in 1673, 1686, and 1687. The French and Indian wars displaced most of the settlers.

When the Massachusetts–New Hampshire boundary was surveyed in 1741, it cut off from Northfield a tract four miles and 197 rods from the northern end of the town. This was the first territory of Vermont granted and owned by English settlers. It was held to belong to the settlers living on it and to be under the jurisdiction of New Hampshire.

4.5.1.4 Maryland

Maryland was the fourth colony established in Colonial America, based on a royal grant. Founders included George Calvert, Lord Baltimore, and other colonists who formulated the land system in Maryland. Colonial Maryland was a "proprietary" colony in that the land of the colony itself was granted to a Lord Proprietor, who then sold or otherwise conveyed the land to others. This land was not truly owned by others in the sense of ownership today but was held in "common socage," meaning that it was still subject to rent and other fines paid to the Lord Proprietor. The current system, known as "fee simple," was established in 1787, allowing for land and property taxes.

From 1633 to 1683 the founders issued land grants known as headrights, in return for immigrants transporting themselves and others to Maryland.

The guidelines for the headright grants were outlined in several acts known as "Conditions of Plantation." The process included individuals proving their rights in and to land, for which they would be granted a warrant for a specified number of acres. The land was then laid out by survey and a patent issued, granting the land to the patentee. The headright system was abolished in 1680, when, under Charles Calvert, the third Lord Baltimore, the Land Office was created. The original land office records, dating from 1634 to the present, include land patents, surveys, grants, warrants, and proprietary rent rolls. After land was transferred into private ownership, deeds and mortgages were filed with the county clerk. A unique characteristic of Maryland land tracts is that each tract surveyed is given a name. Deeds, tax records, and wills will refer to land tracts by name but do not always provide a complete description of the tract. Later surveys may alter the bounds of a tract and often assign the tract a new name, but there are land tracts that are still officially known by names provided during the colonial period.

During the Revolutionary War, Maryland offered land grant lots as a bounty to entice recruits to fill Maryland's enlistment quotas. These lots were located "Westward of Fort Cumberland" in present-day Alleghany and Garrett counties.

4.5.1.5 Rhode Island

Rhode Island was the fifth colony established in Colonial America. The Rhode Island Colony was founded by Roger Williams and other colonists. The earliest land grants were made by the Colony. During its colonial period, the Rhode Island general assembly granted land to groups of colonists who became the proprietors of towns. Beginning in 1659, the proprietors then divided the land into portions, which were granted to individuals along with portions owned in common. The earliest grants were made in the Narragansett area in southern Rhode Island. The records of these grants are known as land evidences.

4.5.1.6 Connecticut

Connecticut was the sixth colony established in Colonial America. Founded by Thomas Hooker and other colonists at Hartford, land was granted by the Connecticut General Court, mostly to groups of proprietors. Later deeds, mortgages, and related instruments were recorded by the town clerk of the respective town in which the land lay.

The area known as the Connecticut Western Reserve was a portion of land claimed by the Colony of Connecticut, mostly in what is now a portion of the state of Ohio. The Reserve had been granted to the Colony by King Charles II. Following the American Revolutionary War, Connecticut, like several other states, gave up claim to some of its western lands but initially sold the Western Reserve to developers. It finally ceded control of this portion to the United

States in 1786 and the area was organized under the Northwest Territory until Ohio was admitted as a state.

Although forced to surrender the Pennsylvania portion (Westmoreland County) of its sea-to-sea land grant following the Yankee-Pennamite Wars and the intercession of the federal government, Connecticut held onto its claim to the lands between the 41° and 42°2′ parallels that lie westerly of the Pennsylvania border. The east boundary of the reserve follows a true meridian along Ellicott's Line, the boundary with Pennsylvania. The west boundary veers more than four degrees from a meridian to maintain the 120-mile width, due to convergence.

Retaining the 3,366,921 acres in Ohio, Connecticut sold the land in the Western Reserve to the Connecticut Land Company for $1,200,000. The Land Company was a group of investors who were mostly from Suffield, Connecticut. They planned to divide the land and sell it to settlers from the east, particularly land-hungry New Englanders.

However, Native American title to the Reserve had not been extinguished. Clear title was obtained for land lying east of the Cuyahoga River by the Greenville Treaty in 1795 and land lying west of the river in the Treaty of Fort Industry in 1805. The western end of the reserve included 500,000 acres of land reserved for residents of several New England towns, which had been destroyed by British-set fires during the Revolutionary War.

The next year, the Land Company sent surveyors led by Moses Cleaveland to the Reserve to divide the land into townships. The townships laid out in this survey were five-mile-by-five-mile squares compared to elsewhere in Ohio, where most townships are six miles on each side per the guidelines of the U.S. Land Ordinance of 1785. Cleaveland's team also founded the city of Cleveland, which became the largest city in the region. The territory was originally named "New Connecticut," which was later discarded in favor of "Western Reserve." In 1800, Connecticut finally ceded sovereignty over the Western Reserve. The United States absorbed it into the Northwest Territory, which organized Trumbull County within the boundaries of the Reserve.

4.5.1.7 New Hampshire

New Hampshire was the seventh colony established in Colonial America. Led by Captain John Mason, John Wheelwright, and other colonists, its first permanent settlement was in 1623.

By 1640, New Hampshire's seacoast was divided among four towns or "plantations": Dover, Portsmouth, Exeter, and Hampton. Inhabitants of these towns, along with settlers in southern Maine, chose to be part of Massachusetts for much of the 1600s, becoming a separate province in 1681. During the years 1641–1680, while a part of Massachusetts, it was governed by the neighboring state. Consequently, grants were made in the form of towns. As a result, grants for the coastal communities were originally issued by Massachusetts. These were subsequently divided into smaller towns, after which they were again subdivided into lots according to division, all of which were different.

In 1622, Captain John Mason and Sir Ferdinando Gorges received a patent from the Council for New England for all the territory lying between the Merrimack and Kennebec rivers. In 1629, they divided the grant, Mason taking for his share an area 60 miles deep between the Merrimack and Piscataqua rivers, which he named New Hampshire. This grant was confirmed to him when the Council for New England surrendered its charter in 1635. When it came time to survey this boundary, it was determined that since it went northward up the Piscataqua River and westward up the Merrimack River, the two end points should be connected with a curved line. In addition, when the distances were measured, a nautical, or geographical, mile was used, greatly increasing the total area received by the proprietors. Attempts by Mason's heirs to make good their claims to this land led to lengthy litigation. The inhabitants were finally compelled to recognize the Mason rights, which were sold (1746) by one of Mason's descendants to a group of 12 Portsmouth men, who became known as the Masonian Proprietors. They issued settlement permits and land titles in the undeveloped parts of Mason's grant. Eventually, the 12 proprietors brought in three more and as a group granted townships to groups of proprietors within their boundaries, which in turn were subdivided, many of the townships being rectangular, along with the lots included therein. The Masonian Proprietors retained lots for themselves within each township besides compensating their attorneys with land grants known as Law Lots.

New Hampshire became a royal province in 1741, with Benning Wentworth as governor. In January 1776, it became the first of the British North American colonies to establish a government independent of Great Britain's authority, although it did not declare its independence at the time. Six months later, it became one of the original 13 states that founded the United States of America, and in June 1788, it was the ninth state to ratify the U.S. Constitution, bringing that document into effect. New Hampshire was the first U.S. state to have its own state constitution.

The Treaty of Paris (1783) left an ambiguous boundary between the United States (formerly Great Britain) and Canada (formerly France) at the northerly extremity of New Hampshire. That treaty defined the boundary as follows:

> From the northwest angle of Nova Scotia, to wit, that angle which is formed by a line drawn due north from the source of the St. Croix river to the highlands, along the said highlands which divide those rivers that empty themselves into the St. Lawrence, and those which fall into the Atlantic ocean, to the northwestern most head of the Connecticut river.

Since there were three tributaries to the Connecticut River, there were three possible interpretations of where "the northwestern-most head of the Connecticut River" could be.

Because of the confusion, in 1832, the residents of this area declared themselves a sovereign nation called the United Inhabitants of Indian Stream Territory, sometimes known as the Indian Stream Republic. They wrote a constitution, created their own legislature and court system, issued their own stamps, and established their own militia. The republic existed until 1836, when the voters had decided that they would be under New Hampshire jurisdiction. The dispute was finally settled by the Webster-Ashburton Treaty in 1842, defining the boundary and incorporating this area as part of the United States.

A large portion of the area now within the state of Vermont was originally considered as within the Colony of New Hampshire. The area now within that state was first permanently settled by European settlers when William Dummer, acting governor of the Province of Massachusetts Bay, ordered the construction of a fort in the approximate location of what is now Brattleboro. Except where influenced by topography, townships in the Vermont territory tended to be rectangular.

In 1741, Benning Wentworth was appointed the first governor of New Hampshire. Benning's grants were usually six-mile squares, which were then subdivided among the proprietors, and six of the lots in each were set aside, one for the Society for the Propagation of the Gospel in Foreign Parts (a missionary organization of the Church of England), one for the Church of England, one for the first clergyman to settle in the township, one for a school, and two for Wentworth himself. Early survey maps of the various townships depict a 500-acre square in one corner of the township labeled "B.W."

The King ruled in 1764 that the New Hampshire Grants in Vermont were to be part of New York, which produced clouded land titles. To obtain clear title, towns petitioned the Governor to give them a new title. One such town, Brattleboro, was surveyed and a plan was produced by Alexander Colden, the surveyor-general of the Province of New York. The map is more detailed and precise than the 1753 New Hampshire Charter map, and it subdivided a large part of the town. It contains a 5400-acre tract divided into 70 lots and depicts three lots of 350 acres each reserved for religious purposes and a 100-acre school lot, all set aside by the Province to encourage the establishment of a local church and school and thus were not conveyed to the grantees. Within the 70 lot subdivision are shown strips of land between tiers and rows of lots, indicating land set aside for future roads, providing access to the lots.

4.5.1.8 Delaware

Delaware, founded by Peter Minuit and New Sweden Company, was the eighth colony established in Colonial America. All the land in the Delaware region had been included in the grant of King James I of England to the Virginia Company in 1606 as well as in the Calvert grant of Maryland by

Charles I in 1632. Such granting of land in America without regard to previous grants was a common practice. That practice gave rise to strife and disagreement during the next 130 years, between Calvert and the Dutch as well as William Penn.

Although claimed by the English based on John Cabot's explorations, the Dutch also claimed the area based on Henry Hudson's journeys and were first Europeans to occupy the area with a trading post near Lewes established by the Dutch West India Company. But after Peter Minuet, who had been the Dutch governor of New Netherland, was discharged by that company, he established the first permanent European settlement with the Swedish West India Company on the Delaware River at the current site of Wilmington. That colony was established in 1638.

The Swedes settled on scattered farms, paying little to no attention to exact boundaries until Dutch law applied making precise property definition important. The Dutch were more organized than the Swedes in establishing their lands. When the Dutch reoccupied in 1659, they began laying out parallel longlots, resulting in each receiving waterfront, meadow, and upland. Longlots continued with Dutch expansion, not only along water bodies but also along roads and paths, and continued during the English proprietary period. Eventually, the Dutch drove the Swedes from Delaware, followed by the English acquisition of the area in 1664.

After the English acquired the area, land grants were issued by the proprietary of James, Duke of York. In addition to all the land in Pennsylvania, William Penn also received, along with a partnership, a grant to land in Delaware in 1682.

An interesting twist to the long history of claims and settlements of this region is what became known as the Delaware Wedge (Figure 4.11), a small tract of land along the common borders of Delaware, Maryland, and Pennsylvania. The ownership of this tract was disputed until 1921 and is now recognized as part of Delaware. Due to both surveying and title concerns, the nature of the various grants and their associated boundaries left this tract in question as to which state it was part of, for many years. The tract is bounded on the north and west by a circular arc, known as the New Castle Twelve-Mile Circle, with a 12-mile radius centered on the cupola of the courthouse at Newcastle, Delaware. The arc runs from an eastern extension of Mason-Dixon line to the 1934 mean low water line of the Delaware River.

4.5.1.9 North Carolina

North Carolina was the ninth colony established in Colonial America. The North Carolina Colony was founded by the Virginia Colonists. Various royal, colonial, state, and federal governments made the first claims to land in what is now North Carolina. Much of this land was given or sold to individuals. North Carolina went through an evolution of land granting. The first period was the Provincial or Proprietary Era, which extended from 1663

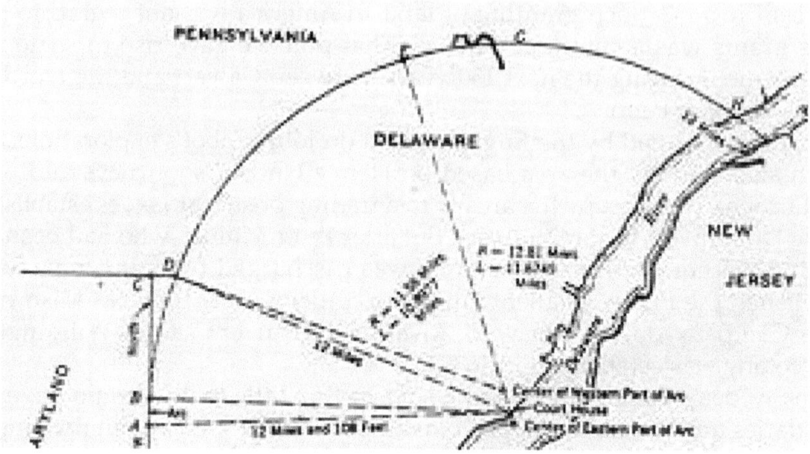

FIGURE 4.11
New Castle 12-mile circle. (From *Delaware State Boundaries* by William Schenck, Delaware Geological Survey Information Series 6, University of Delaware.)

to 1729. The Charter of Carolina was issued in 1663 by King Charles II as a grant in America to eight men, whom he selected. They were known as Lords Proprietors, or simply Proprietors, and were given very large amounts of land. The lands granted included all the land from the southern border of the Virginia Colony to Florida. It also granted all the land between these northerly and southerly borders, from the Atlantic Ocean westward to the Pacific Ocean (from sea to sea). During the time the proprietors controlled the land, North Carolina was a proprietary colony.

The next period (known as the Colonial, British Crown, or Royal Era) extended from 1729 to 1775. By 1729, seven of the eight proprietors had sold their shares to King George II, which made North Carolina a Royal Colony. Most of the land was located in the southern half of the state. The one proprietor who did not sell was Lord Granville, whose agents made grants to vacant lands. In 1737, Henry McCulloh, later known as the largest land speculator in colonial North Carolina, received a royal patent for 1.2 million acres in western North Carolina. The lands granted by these latter two holders were technically deeds, not grants.

During the Revolutionary War, North Carolina created a land office in every county, each having its own appointed entry-taker and surveyor. As early as 1735, many North Carolina patents were issued by the Crown, which were later discovered to be part of South Carolina. The boundary between the two states was not officially drawn until 1772. After that survey, lands formerly granted by North Carolina were registered in South Carolina and are often called North Patents. South Carolina issued new grants for some of them.

The area currently within the state of Tennessee was originally part of North Carolina. North Carolina received permission from the federal government to continue to finalize the military warrants as well as to finalize all other types of entries initiated before 1790. In 1791, the U.S. Congress requested North Carolina to provide a list of all grants, including military grants, issued for land in what later became the state of Tennessee. The list totaled 5486 grants.

Although Tennessee is considered a metes and bounds state, a large portion of it was also set apart in townships and ranges. Tennessee was unable to grant its own lands for the first 10 years, until 1806 when Tennessee assumed the granting of lands within its territory. The southeast portion belonged to the Cherokee Tribe and the western part belonged to the Chickasaws until 1819 and the south central part was originally reserved for Congressional use and was known as the Congressional Reservation. The area open for settlement was divided into surveyor districts, setting up 7 of what eventually became 13 surveyor's districts, thereby dividing the state into ranges and sections similar to the system in the Northwest Ordinance. This was Tennessee's attempt to move away from what was termed "the chaotic metes and bounds system" inherited from North Carolina and Virginia. However, there was such a strong precedent for using metes and bounds that even though most of the Surveyor Districts were in fact surveyed in a grid, and land was granted according to named lots, the section/range system quickly fell into disuse. Deeds continued to be written with metes and bounds descriptions, and today, it is somewhat difficult to determine where the old section lines were originally established.

Today, a few areas east of the Tennessee River are mapped by range-township; the rest of the state is mapped by metes and bounds. However, grants, especially the earlier ones, are often either wholly or partially rectilinear.

4.5.1.10 South Carolina

South Carolina was the 10th colony established in Colonial America. The area now in South Carolina was included in the 1663 Carolina grant. Gradually, due to the large size of the colony, the northern and southern regions were administered separately, although they continued as one colony until 1729.

Prior to the surveys of the border east of the Catawba River (1764) and the survey west of the River (1772), over 1000 land grants were issued by North Carolina in territory, which is now South Carolina. Like neighboring states to the north, many grants were either partially or totally rectangular in shape.

4.5.1.11 New Jersey

New Jersey was the 11th colony established in Colonial America. A study of early land grants and tenure in the colony demands consideration of Maryland and Delaware, since the region was claimed and administered by the same factions. Dutch activity came as an extension of the Dutch West India Company from New Amsterdam. Five patroonships were claimed in

New Jersey and Delaware, three along the Hudson River in northeastern New Jersey and one on each side of Delaware Bay. None existed for any length of time. After a British fleet sailed into the area in 1664, the British colonization period began.

New Jersey history during the British colonial period can be divided as follows: the period of the first proprietors, beginning with the grant of King Charles II to the Duke of York and his reconveyance to Lord John Berkeley and Sir George Carteret, in 1664, and extending to 1682; that of the 24 proprietors extending from the latter date to 1702; and the period of the royal province ending with the Revolution.

During the period of multiple proprietors, the colony was divided in two parts—East Jersey and West Jersey. In 1702, the two ownerships turned their control over to Queen Anne, who united the two into a single royal colony. While existing as East Jersey and West Jersey, proprietorships existed and townships were granted. East Jersey preferred townships with compact settled cores, while West Jersey favored townships as established in New England. A variety of grants were made, from small to large, the smaller on a headright basis. The records of East and West Jersey Proprietors document nearly three and a half centuries of land transactions and settlement in New Jersey.

New England style townships, the first English settlements, mimicked those in New England by distributing lots according to divisions. Friction arose among the settlers, with some receiving lands from original granting, some from proprietors, and others from purchases from the Indians, which sometimes caused conflicts. Most grants in New Jersey were of smaller parcels, all or nearly all surveyed at some point early on.

4.5.1.12 Pennsylvania

Pennsylvania was the 12th colony established in Colonial America. The Pennsylvania Colony was founded by the Quaker William Penn and other colonists. Yet, land grants by the Dutch and Swedish governors and settlements by nationals of various European countries had been made within the present boundaries of Pennsylvania many years before the land in Pennsylvania was granted to William Penn. As examples, Dutch Agent Ardent Corsson obtained a large tract "on and around the Schuykill" by agreement with the Indians and Swede Swen Schulte obtained possession of practically all the land in sections of Philadelphia. When the English drove out the Dutch in 1664, Colonel Richard Nicolls, as English governor, granted land and confirmed some grants that had been made by the Dutch and Swedish governors.

In 1681, Charles II granted all the land in Pennsylvania to William Penn (Figure 4.12). He received, along with his heirs and assigns, the power of feudal lords, with authority of a viceroy. He sent his cousin, William Markham, and his surveyor, Thomas Holmes, to Pennsylvania to deal with boundary issues. They arrived July 1, 1681, and immediately began selling lands in

> Charles the Second, by the Grace of God, King of England, Scotland, France, and Ireland . . . Doe give and Grant unto the said William Penn, his Heires and Assignes, all that Tract or Parte of Land in America, with all the Islands therein contcyned, as the same is bounded on the East by Delaware River, from twelve miles distance Northwards of New Castle Towne unto the three and fortieth degree of Northerne Latitude, if the said River doeth extende so farre Northwards; But if the said River shall not extend soe farre Northward, then by the said River soe farr as it doth extend; and from the head of the said River, the Easterne Bounds are to bee determined by a Meridian Line, to bee drawne from the head of the said River, unto the said three and fortieth Degree. The said Lands to extend westwards five degrees in longitude, to bee computed from the said Easterne Bounds; and the said Lands to bee bounded on the North by the beginning of the three and fortieth degree of Northern Latitude, and on the South by a Circle drawne at twelve miles distance from New Castle Northward and Westward unto the beginning of the fortieth degree of Northern Latitude, and then by a streight Line Westward to the Limitt of Longitude above-mentioned.

FIGURE 4.12
Transcript of Section 1 of the Royal Charter of King Charles II of England to William Penn of March 4, 1681.

Pennsylvania. The oldest deed known is in Bucks County, dated April 1, 1681, from Penn to Thomas Woolrich for a thousand acres. Based on correspondence with Holmes, during the first 14 months, 565,500 acres was sold in parcels of from 250 to 10,000 acres. Penn followed the examples of the neighboring provinces in granting land, along with his personal experience in previous grants in New Jersey. In every 100,000 acres surveyed, Penn reserved 10,000 acres for himself.

Arriving on the American shore on October 24, 1682, Penn went ashore at New Castle, where he announced his proprietorship of Delaware. In December, he went to Maryland to see Lord Calvert, who had disputed the exact location of the southern boundary of the Penn grant, beginning a long history of disputes between families in Pennsylvania, Delaware, and Maryland. Besides the 40° latitude boundary (Figure 4.11), his grant provided that this parallel was to be intercepted on the west and northwest by a circle drawn 12 miles distant from New Castle. The dispute was due to a lack of geographical knowledge in London at the time of drafting the Penn grant. Calvert had already marked the 40° parallel by a fort on the Susquehanna River, 20 miles north of the present Maryland boundary. Penn's political position enabled him to prevail in his interpretation of the boundary and to push his southern boundary 20 miles south of the Susquehanna fort, thereby supposedly taking from Calvert more than 3 million acres along the southern part of Pennsylvania.

At the time of Penn's arrival, there were 5000 Europeans living along the Delaware River, including English, Swedes, Finns, Danes, Dutch, Irish,

Scotch, and French, with the number of English being equal to all the others combined. In addition to his grants, Penn made seven purchases from the Indians. He wrote, "I have followed exactly the counsel of the Bishop of London, by buying, and not taking, the native's land." He located a manor of 6500 acres in Bucks County, where he built and furnished an elaborate residence. Several tracts of 10,000 acres each were laid out as manors for selected relatives.

Penn returned to England for 15 years before again setting foot on American soil. He continued to grant lands until his death in 1718. At the time of his death, he held title to more than 27 million acres. His will provided lands to his widow and other trustees, with instructions for conveyances to his children by his first wife. The remaining land was designated to go to the children of his second wife. As recently as 1954, in the Pennsylvania case of *Bruker v. Burgess and Town Council of Borough of Carlisle*, a claim on a parcel of common land dating back to the heirs of William Penn was upheld by the courts as being valid.

Another major land owner in Pennsylvania was William Bingham, a lawyer and banker who owned a sizeable estate in Philadelphia. Besides Maine (then Massachusetts) and New York, he acquired large tracts of land in Pennsylvania. He had valuable property in the city and county of Philadelphia, and from the Commonwealth of Pennsylvania, he purchased over a million acres. His title to that land lying chiefly in the northern counties (Tioga, Bradford, and Potter). In Tioga county alone, he owned 229,343 acres founded upon warrants granted to him by the Commonwealth in the years 1792 and 1793, upon surveys made by the officers of the Commonwealth and regularly returned by them into the office of the surveyor-general, and upon patents issued to him or to the Trustees appointed in his will and signed by the Governor or Secretary of the Land Office.

4.5.1.13 Georgia

Georgia was the last of the 13 original colonies to be established. The state is unique in that although it was one of the original colonies, the majority of the state's current land area was originally subdivided with a rectangular cadastral survey.

Georgia was established in 1732 as the 13th American colony based on a grant from King George II to James Oglethorpe and other colonists. William Bull, surveyor from South Carolina, along with Oglethorpe, undertook the first surveying project in Georgia—the laying out of the Town of Savannah. The town plan consisted of a series of wards, each subdivided into squares, tythings, and lots. Each ward was planned around a central square, which was bounded on its eastern and western sides by four lots called trust lots, which were reserved for public uses, and on the north and south by four tythings, each divided into 10 house lots, of which each fronted on a street and backed on an alley.

In 1737, Surveyor Hugh Ross was commissioned to lay out the lots for the town of Ebenezer about 25 miles northwest of Savannah. (That town site was later abandoned and is now a ghost town.) Because of the less than desirable arrangement of farm lots laid out in Savannah, standards for lot layout there were relaxed, and even though the farm lots were still laid out in contiguous, rectangular tracts, Ross was instructed to lay them out clustered around rich creek bottomlands, avoiding pinelands as much as possible. As a result, the lots were long, narrow strips, many fronting on Ebenezer Creek, resembling the river lots of the Northeast and the French arpent system of long narrow lots fronting on bodies of water, particularly rivers. Other lots in Ebenezer were rectilinear.

In 1754, King George II established a civil government with a governor and surveyors-general. Finding the system of land ownership and disposition in a state of confusion, the governor ordered new titles to be issued and abandoned the previous system of land granting in favor of a headright system overseen by the surveyors-general. Once awarded a headright, grantees could select any available land at a location and configuration of their choosing. After the American Revolution, the State passed a land act in 1783 that continued the headright system. That system provided settlers the right to 200-acre grants as a headright plus 50 acres per dependent. The land was offered at no cost for the first 200 acres and 1 to 4 shillings per acre thereafter, with a 1000-acre maximum.

In 1784, two counties instructed surveyors to lay out lands into square or oblong shapes. Towns and cities were laid out almost entirely with lots and blocks, with streets between them. For the most part, wherever possible, in subdivisions of towns and grants, the rectangular concept prevailed. Unfortunately, in other areas, since grantees could choose where they wanted their lot under both the British and the State headright systems, the land was in disarray, with gaps and gores between and overlaps of selected parcels. Further, there were frauds and scandals associated with the program, including the infamous Yazoo Land Fraud. Fortunately, most of the headright grants were confined to the coastal region of the state. The remaining three quarters of the present area of the state was subdivided and granted under a rectangular cadastral survey associated with the Georgia Land Lottery.

The cadastral survey associated with the land lottery began when two strips of land were ceded to Georgia by the Creek Native American tribe in 1802 in return for lands west of the Mississippi. Because of the problems associated with the headright system, it was decided to distribute the new land in an improved system. Three counties (Baldwin, Wayne, and Wilkinson) were created from the ceded tracts, one of which was in the southeast corner of present-day Georgia and one just west of the Oconee River, and each county was divided into several land districts. For each of the 13 resulting districts, a surveyor was hired to map and segment the land into lots, which were laid out in a rectangular grid pattern, ensuring the matching of adjacent boundaries. Each of the lots in Wayne County were 490 acres and each of the lots in Baldwin and Wilkinson Counties were 202 1/2 acres in size. The surveyors were paid $2.75 per mile, which had to include all incidental costs, such as field notebooks and salaries for anyone who they hired to assist them. Once the survey of the district was complete, the surveyors provided their records to the Georgia surveyor-general.

To ensure a fair distribution of the land, a lottery system was used. Participants in the lottery, which was held in 1805, had to be citizens of the United States and to have resided in Georgia for at least a year and to be one of the following: bachelors 21 years or older, married men, widows with a child under 21 years of age, or orphans under 21 years of age. When the total number of applicants was known, tickets for each of the lots for distribution were placed in a rotating barrel. Blank tickets were added so that the number of tickets was the same as the number of applicants. As each name on the list of eligible individuals was called, commissioners appointed by the governor would draw a ticket from the barrel. The individual received either a blank, which indicated no prize, or the opportunity to purchase the lot shown on the ticket for a cost of $8.10 for the 202 1/2-acre lots or $19.60 for the 490-acre lots. All of the 4580 whole lots were distributed in this manner during the 1805 lottery, while any fractional lots or islands less than 100 acres were held out and sold at public auction in 1806.

Within 25 years after the first cession of land by the Creek tribe, the remainder of the Creek lands as well as land of the Cherokee Nation had been ceded, in some cases forcibly, to the State. That land was subdivided into rectangular lots in a manner similar to the surveys associated with the 1805 lottery and distributed by seven different additional lotteries in 1807, 1820, 1821, 1827, 1832, and 1833, wherein lotting was almost exclusively rectangular.

The difference between the Georgia cadastral system and the U.S. Public Land Survey is that the former has no coordinate system of meridians and baselines. Also, unlike the federal system, no standard system of numbering was specified by the Georgia land lottery acts until 1830. Rather, the Georgia system began with "boxes" (counties), created smaller "boxes" (districts) within them, and then created still smaller "boxes" (land lots) within these. The lots are numbered, not according to a standard numbering system. Also, instead of being oriented to the cardinal directions, the lots are oriented either 45° or 60° from the meridian or, as in the case of original Wayne County, 13° from the meridian (Cadle 1991).

The cadastral survey associated with Georgia's lottery system was the most extensive of all of the colonies and the largest ever held in the United States (Hilliard 1973). Lots were surveyed in five different sizes based on the perceived quality of land, which was a concept employed in one form or another in many of the rectangular systems of the colonial states. In the 1805 lottery, as previously mentioned, lots were 202.5 acres and 490 acres. In 1807, lots were 202.5 acres. In 1820, the lots were 250 and 490 acres. In 1821, the lots were again 202.5 acres. There were two lotteries in 1832, the regular lottery, wherein lots were 160 acres, and the 1832 Gold Lottery area, lots were 40 acres in size. The total area distributed in that manner represented well over three-fourths of the current State of Georgia (Figure 4.13). District maps, field books, and plat books for the land lotteries have been maintained by the Georgia Archives. Since these surveys represent the original subdivision of the land, the conveyances by the lotteries are important as the root of title for the chain of title for this land today.

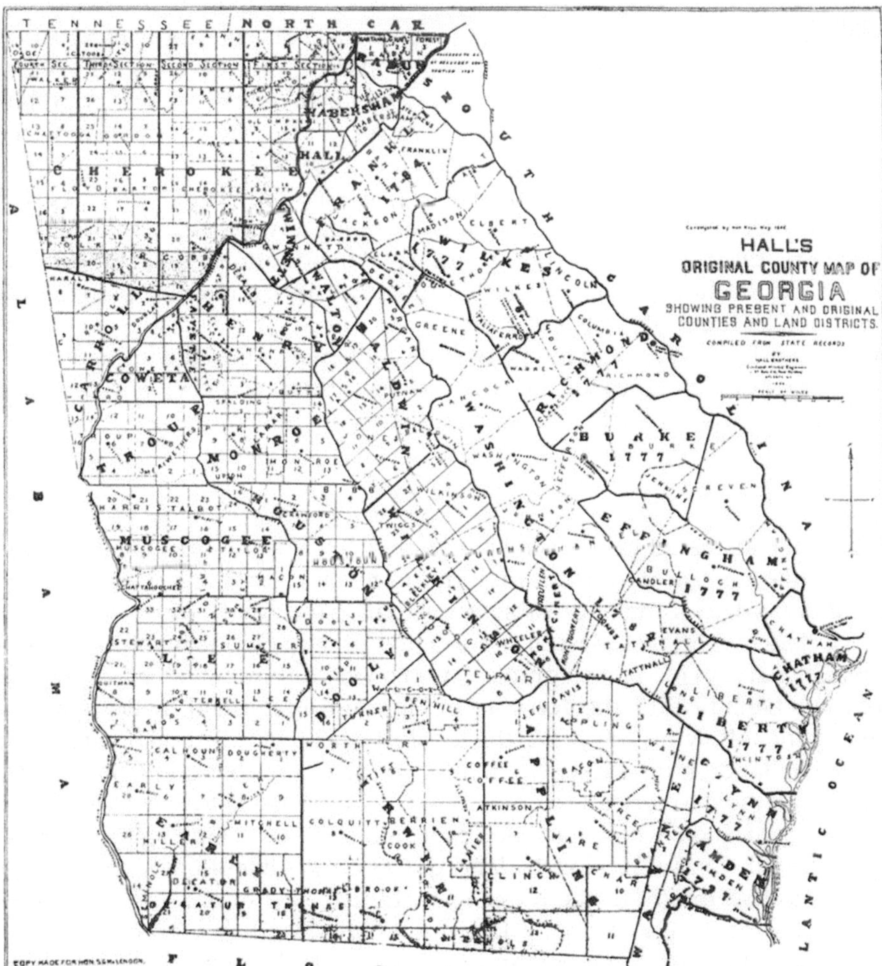

FIGURE 4.13
Hall's 1895 County Map of Georgia. Note that while lands in the eastern areas were primarily subdivided under the headright system, areas with rectangular tracts, which represent area subdivided into smaller rectangular lots under the Georgia Land Lottery system. (From Georgia Archives, University of Georgia.)

There were several isolated tracts of land in Georgia that were not laid out under either the headright system or the land lot and district system. These are known as the Reserves. Twelve such reserves were laid out in rectangular fashion, the number of lots being dictated by the size of the reserve. Those with cities, such as Columbus and Macon, were laid out as typical cities, with lots, blocks, and streets.

Another unique feature regarding boundaries in Georgia is the widespread use of circular political boundaries, where the boundaries of communities are, in part or in their entirety, circular in shape. There are dozens, if not hundreds, of such boundaries in Georgia. This concept is not limited to Georgia, however, as similar towns are also found in the Carolinas and Alabama. Nevertheless, Georgia contains by far the most circular boundaries. It is an interesting question as to the reason for the prevalence of circular boundaries in the area since they are typically not common except in Georgia and the surrounding states. The *Atlas of Georgia* (Hodler 1986) suggests that circles were used because of "...the advantages of explicit verbal clarity, directional impartiality, and ease of adoption." Further, circular boundaries do have ancient origins. In Plato's Laws, a circular plan is proposed for an ideal city. Also, the Roman architect Vitruvius suggested that the circle is the ideal shape for a city (although, in practice, the Romans used the concept of rectangles for the subdivision of their territory). Furthermore, several historic sites, such as Stonehenge, are circles.

4.5.2 Florida

Although considered to be a public land state, Florida has areas with unique rectangular cadastral systems in addition to the U.S. Public Land Survey. Therefore, a discussion of those systems will be included.

One class of original cadastral surveys prevalent in Florida involves the confirmation surveys of Spanish grants. Based on its "discovery" by explorer Juan Ponce de Leon in 1513, the Territory of Florida was claimed by the Spanish. During the following 250 years, the territory was controlled by Spain, and a number of land grants were issued. In 1763, after numerous conflicts with the British and their American colonies, Spain traded Florida to Britain in return for control of Cuba, which had been captured by Britain. During the 20 years under British control of Florida, land grants were given to officers and soldiers who had fought in the French and Indian War in order to encourage more settlement. In addition, a large number of settlers moved to Florida from other British colonies in America. After the American Revolutionary War, as part of the 1783 Treaty of Paris ending that war, Britain was required to return Florida to Spanish control. Spain then began to encourage further settlement in the territory by offering headright grants of 100 acres plus 50 acres for each family member, including slaves, to any settler who would take an oath of allegiance and were self-supporting. The grants required a 10-year probation period during which the land could not be sold and during which the land had to be cultivated, a house and fences built, and a certain number of livestock maintained. Thus, there were numerous settlers in Florida prior to it becoming part of the United States.

In 1821, Spain ceded the Florida territory to the United States after years of troubled relations between the two nations. This included frequent incursions into Spanish territory by the U.S. Army in pursuit of Seminole Indians who had begun raiding Georgia settlements and also offering havens for runaway slaves. To address the issue of existing land claims, Article III of the Treaty of Acquisition called for all Spanish grants to be ratified and confirmed to the persons in possession of the land at the time of the treaty. To implement this, in 1822, Congress created a three-person Board of Commissioners who established a process for confirmation of such Spanish grants as well as earlier British grants that had been validated by the Spanish Government and also of claims by persons, such as squatters, who were occupying land in Florida prior to the treaty without an official grant. This process is in accord with a core principle of American jurisprudence when acquiring jurisdiction over territory from other sovereign nations that legitimate individual landowners from a previous regime should be afforded protection for their private property rights.

The grants of land preceding acquisition of Florida by the United States were typically rectangular in shape, except when located abutting a watercourse. Surveys of the grants were usually made during the process of acquiring the grants from Spain or Britain. In addition, surveys were made to exclude the grants from the public land during the course of the U.S. Public Land Survey following acquisition of the territory by the United States. Those surveys represent important documentation of the root of title for large portions of Florida. The surveys associated with the grants are in a series of dossiers associated with each confirmed grant in the state's general archives. The surveys made of the grants excluding them from the U.S. Public Land Survey are in a series of bound books with a survey sketch and metes and bounds descriptions of each grant located in the State Lands Office. Unfortunately, most of the records for West Florida are missing (Cole 1990).

Other important original cadastral surveys in Florida are those associated with the Forbes Purchase. Sometimes referred to as "Florida's Biggest Land Grab," activities over a period of more than 52 years (1783–1835) and over a million acres of land resulted in a fascinating history involving land titles and surveying, much of which became far-reaching This purchase and land grant is one of the largest in American history.

The history of this grant began in 1765, when a Scottish immigrant named John Panton sailed to America in 1765 and settled in Savannah, Georgia. By 1774, he had become a partner in the trading firm of Moore & Panton, gaining invaluable experience as an Indian trader. The following year, he was appointed by the British as the official trader to the Creek Nation. Forming a new company with another Scotsman named Thomas Forbes; they called themselves Panton, Forbes & Company. The two were Tories and, remaining loyal to Britain, relocated their headquarters to St. Augustine in

Spanish-controlled Florida when the Revolution began. Their properties in South Carolina and Georgia were confiscated by the United States, which created animosity. Nevertheless, the company continued to expand in Spanish Florida throughout the Revolutionary War.

The pair was later joined by three other Scotsmen, forming Panton, Leslie & Company in 1783, the year that the British Government issued their chartering paperwork. Prior to this, during the war, the company became well established. The Spanish, realizing that there were no experienced Spanish Indian traders, gave the company a monopoly in East Florida. In 1785, the company expanded even more and moved its headquarters to Pensacola; Forbes' and Leslies' younger brothers joined the company; and John Forbes opened a post in Mobile, Alabama. Within 10 years, the company had trading posts as far north as Memphis, westward to New Orleans, all along the Gulf Coast into Central Florida, as well as the Bahamas and other islands in the Caribbean. During all of this time, the company established a working relationship with the Creeks, Seminoles, Chickasaws, Choctaws, and Cherokees. By the late 1790s, the company had acquired 3 million acres of land in what later became the states of Mississippi and Alabama.

As the company expanded, it extended credit to its Native American trading partners and developed a large accounts receivable balance from sales on credit. In 1804, to pay off that debt as well as for compensation for losses in two robberies by Native Americans of their trading post in St. Marks, the Creek and Seminole Native American Tribes ceded to the company a large tract of land in North Florida that had been recognized by the Spanish as being owned by the tribes.

The deed of cession, which was approved by the Spanish governor at New Orleans in response to a petition by the Company, was signed by 24 Lower Creek and Seminole chiefs on May 25, 1804, at Cheskatalafa, an Indian village located on the west side of the Chattahoochee River in present-day Henry County, Alabama. The deed included the following language:

> [W]e, the undersigned chiefs of the Seminole tribe assembled together, having maturely weighed the enormous debts which we owe to the house of Panton, Leslie & Co.,… and being likewise responsible for the robberies and depredations which, on two occasions, we have perpetrated on the stores of the aforesaid house… we have determined… to sell and to cede… a district of land which we hold as actual owners and proprietors.

Seven years later, after the debts had piled up again, three additional large tracts were transferred to the company, which had changed its name to John Forbes and Company. The total of all the grants was approximately 1.4 million acres and included all of the present Florida counties of Liberty and Franklin as well as large portions of Gadsden, Leon, and Wakulla counties.

The Company authorized a survey of the land by Asa Hartsfield of South Carolina in 1807. His first charge was to survey the northern boundary of the entire tract, which had been marked by the Native Americans in 1804. The second task was to inspect the entire tract and determine areas best suited for settlement and cultivation since much of the tract was wetlands in character. He identified only two such areas, one lying along the northern boundary and one in the southeastern corner. The latter tract, covering 37,000 acres, was then subdivided with what is known as the Hartsfield Survey. The lots in that survey were 360 acres in size, about 3/4 miles on each side. Unlike the United States Public Land Survey (USPLS) survey, the lots were oriented at a bearing of N17 1/2°W, apparently to be perpendicular to the general trend of the shorelines of the Gulf of Mexico and the St. Marks River.

In 1819, the Forbes Company sold the total tract to Colin Mitchel, a Spanish merchant, for $135,000 (a price of about nine cents per acre), who intended to further subdivide and sell the land. In 1821, Mitchel contracted Surveyors Brown and McBride to survey the other area identified by Hartsfield as ideal for development. That survey, known as the Little River Survey, subdivided the 28,460-acre tract into 800-acre lots, each being 100 chains in length on the east–west sides and 80 chains on the north–south lines and oriented at a bearing of N7 1/2°E. The reason for the orientation of this survey is unclear.

After the territory of Florida was acquired by the United States in 1821, the private ownership of the Forbes Purchase tract was challenged by the United States during the grant confirmation process (see previous section). The challenge was based on the well-established policy of not recognizing land transactions between Native Americans and private individuals. (See Section 3.1.7 regarding Aboriginal Law.) But the Board of Commissioners disbanded and left it to Congress to decide whether the Forbes Purchase was valid, and Congress also failed to act. Ignored for seven years, Mitchel filed suit in the territorial court, where a judge ruled against the validity of the Forbes Purchase. In 1831, Mitchel filed an appeal to the U.S. Supreme Court. As a result, the dispute was finally settled by that case (*Mitchel v. United States*) in 1835. In that case, the United States argued that, based on the case of *Johnson v. M'Intosh,* the European "discovery" of America automatically divested native Indians of the "power to dispose of the soil at their own will, to whomsoever they pleased." Nevertheless, the court ruled that what distinguishes *Mitchel* from *Johnson* is the fact that the Spanish government both approved and confirmed the Forbes Purchase. In the opinion, the court specifically mentioned the "large amount due by the Seminole Indians" and "robberies of their stores in 1792 and 1800 by members of that tribe, headed by the celebrated adventurer Bowles, exceeding in all $60,000, of which they were unable to procure any payment from the Indians, but who had expressed a willingness to make compensation by a grant of their lands." The entire process, emanating from a European claim at the time, illustrates the importance of having an understanding of the origin of rights and titles.

This court order in the case also directed the surveyor-general of Florida to survey and lay off the lands described in the conveyance. As a result, the entire boundary was run with a lengthy traverse around the perimeter, including the islands included in the Purchase, and tied to the USPLS grid. A large portion of the Purchase had already been subdivided into townships and ranges by the Public Lands Survey between 1824 and 1826 before the legal dispute began. Interestingly, even though the land was not part of the U.S. public domain, the interior portion of the Forbes Purchase not covered by the Hartsfield and Little River surveys has also been subdivided using the conventional USPLS township and range system. This was accomplished by a combination of privately contracted surveyors and U.S. Public Land Survey surveyors. For this work, no contracts can be found, so it not clear whether or not this was authorized by the federal government.

As a result of this complex history, surveying in the Forbes Purchase area can be challenging (Figure 4.14). A plat and field notes are available for the Hartsfield Survey, but the plat does not have measurements and the field notes do not mention any corners set. No field notes can be found for the Little River Survey, although a plat, without measurements, is available. Field notes and plats with a few details are available for the areas subdivided into conventional townships, ranges, and sections. Fortunately, most of the area is sparsely populated, which has minimized the impact of the poorly documented original surveys.

4.5.3 Hawaii

Hawaii is generally considered to be a metes and bounds state. Nevertheless, when titles were originally created from the King, an orderly design was established upon the landscape. It was not strictly a rectangular design, but rather was strongly influenced by the shape and topography of the individual island being divided and the culture of the island. Therefore, understanding land ownership, tenure, and boundaries in Hawaii requires a study of the history of Hawaiian culture itself.

Ancient Hawaiians had no concept of private property, but they did follow a complex system of land division. All land was controlled by the highest chief or king, who held it in trust for the entire population. Supervision of the lands was designated by the king based on rank or standing. An entire island, known as an island kingdom, or *mokupuni*, was divided into smaller units, down to a basic unit belonging to a single family. Each *mokupuni* was divided into several *moku*, the largest units within each island, usually wedge-shaped, and running from the sea (*makai*) inland to the mountain crest (*mauka*). The island of O'ahu, for instance, was divided into six *moku* (Figure 4.15).

Each *moku* was divided into *ahupua'a*, narrow wedge-shaped land sections that again extended from the mountains to the sea. The *ahupua'a* was shaped by island geography and was a wedge-shaped area of land running from the uplands to the sea, following the natural boundaries of its watershed. Each

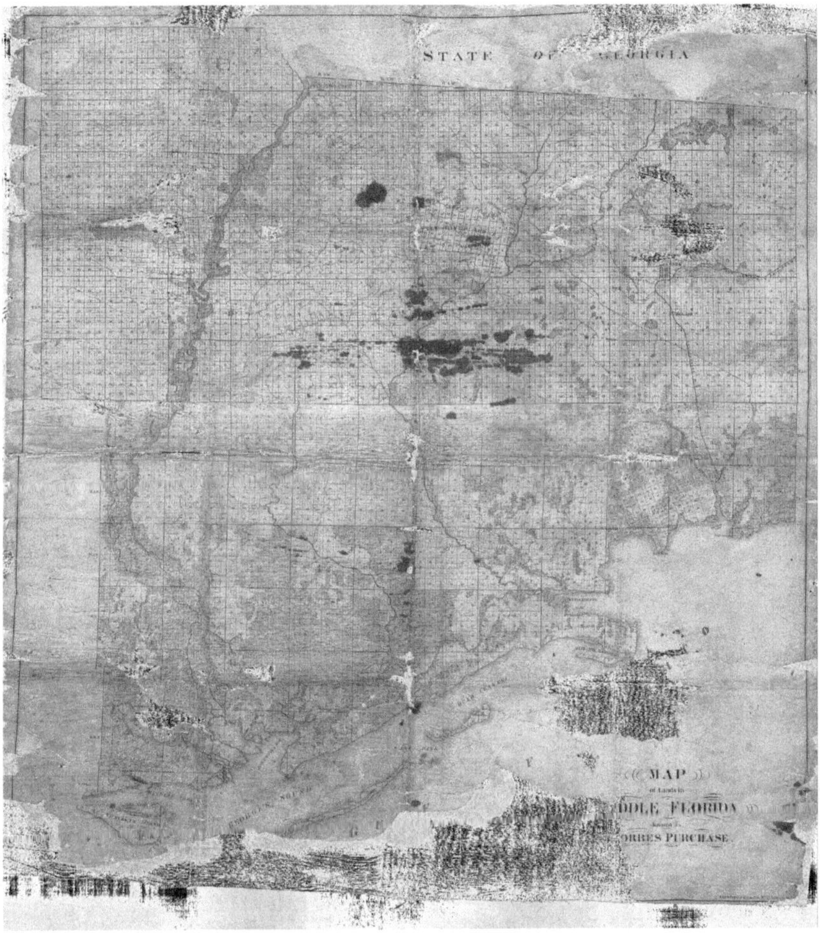

FIGURE 4.14
Map of Middle Florida showing the Forbes Purchase and surrounding lands. Note the Hartsfield Survey in the southeast corner and the Little River Survey in the northwestern portion of the Purchase.

one contained the resources needed by the human community, from fish and salt, to fertile land for farming taro or sweet potato, to koa and other trees growing on the upslope areas. Generally, the poorer agricultural regions were split into larger *ahupua'a* to compensate for the relative lack of natural abundance. Each *ahupua'a* was ruled by an *ali'i*, or a local chief, and administered by a *konohiki*. The word *ahupua'a* itself comes from the combination of the words *ahu*, or stone altar, and *pua'a*, which means pig. At the boundary between *ahupua'a*, a stone altar, known as a *heiau*, was built and a pig's head was carved out of wood, placed on top. This type of monument was not only to mark the boundary but also was a place to bring the offerings for the *ali'i*, *ho'okupu*, once a year.

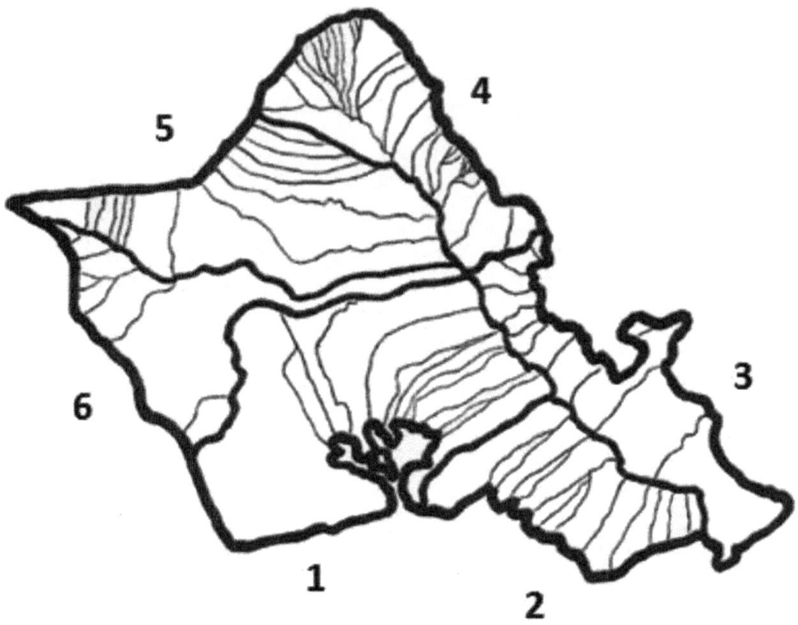

FIGURE 4.15
Diagram of subdivision of land in Hawaii.

Within an *ahupua'a* were two or three smaller divisions known as *'ili* that comprised the estate of the chief. Each *'ili* could consist of noncontiguous pieces called *lele*. *Mo'o* were sections of the *'ili* that were arable and usually did not extend to the sea. Smaller yet were the *kuleana*, like the size of *ahupua'a*, which depended on the natural fertility and abundance of the land.

Although no private ownership of property existed, land tenure of the *maka'ainana* (commoners) was stable. They paid weekly labor taxes and annual taxes to the *konohiki*, or local overseer, who collected goods to support the chief and his court. The *konohiki* supervised communal labor within the *ahupua'a* and also regulated land, water, and ocean use.

As early as 1842, government lands began to be set apart, and a treasury board was appointed to administer those lands in which it had an undivided interest and undefined claim in all land in the Kingdom until the Mahele. The great mass of lands consisted of those surrendered and made over to the government by King Kamehameha III, enumerated by name in the Act of 1848. In addition were the lands ceded by the several chiefs, those lands purchased by the government over time, and the lands forfeited to the government by the neglect of their claimants to present their claims by designated dates. A Land Office was created in 1846, authorizing the minister of the interior to sell or lease government lands. From 1850 to 1860, nearly all desirable government land was sold, generally to natives, and surveyed at the expense of the purchaser.

In 1846, an act was passed that included the appointment of a Board of Land Commissioners "to quiet Land Titles." The powers of the Board were extended repeatedly until their dissolution in 1855. They received full power to investigate and finally confirm or reject all land claims arising prior to 1845, with decisions subject only to appeal to the Supreme Court. The Board's rules included a provision that every piece of land should be surveyed at the claimant's expense.

The traditional system ended in 1848, when King Kamehameha III was persuaded by foreigners to institute the Great *Mahele* (division), allowing land to be bought and sold. At this time, most *ahupua'a* were divided. Three on the island of Hawaii are among the few large *ahupua'a* that remained nearly intact under single ownership, except for some kuleana lots, because they were crown lands owned personally by the monarch. In spite of this division, the impact of the original *ahupua'a* boundaries can be seen in many areas today.

To accelerate the process of land distribution, an Act was passed in 1852 to grant titles to Konohiki(s) for whole "Ahupua's(s) or Ili(s) received by them in the Mahele of 1848, by their proper names without survey." The greater part of this class of lands was awarded in this way. By an act passed in 1862, The Commission of Boundaries was established to provide the means of legally settling the boundaries of these lands and of a few that had been patented by name only, "before the testimony of witnesses should be lost by reason of death." This is known as *kamaaina* testimony.

Lands reserved by Kamehameha III in 1848 "for himself, his successor forever" as his private property are termed Crown Lands. In addition to these are a few lots in Honolulu, and Lahaina, awarded to him by the Land Commission. The King and his successors have treated these lands as their private property, selling, leasing, and mortgaging at will.

An act passed in 1850 stated that about "one-twentieth part of all the lands then belonging to the Government should be set aside as School Lands for the general purposes of Education." Shortly after the passage of the Act, certain lands and school sites were designated and set aside for the purpose intended. As later provided, most of the School Lands have been sold. The sales of Government Lands have always been by metes and bounds, and the original surveys and plans placed on file, except for certain *Ahupua'a*(s) sold by the Board of Education, for which grants by name exist.

An act in 1884 was intended to allow the acquisition and settlement of homesteads. The Minister of the Interior was authorized to survey and lay out Homestead Lots from the public lands, not to exceed 20 and not less than 2 acres, in dry or kula land, and not over two acres in taro land, with convenient roads in connection therewith. After advertisement, these lots were assigned to the first applicant with conditions for the use and settlement. In 1888, the Act was amended so that in rocky districts of Kahikinui and Kipahulu, Maui, and Kona and Puna, Hawai'I, the size limit of the Kula lots was increased to 100 acres.

There were unassigned lands, most on the island of Hawai'i, which had been overlooked in the "Mahele" of 1848 and for which no title existed. All private claims had to have been brought forward by certain dates, along with claims under the Mahele Book not presented before 1862, by law reverted to the government. The question remained whether they belong to the class of government lands or to that of Crown Lands or to the lineal heirs of Kamehameha III. This question has since been settled by the Supreme Court in favor of the government.

According to Alexander (1882), the early surveys, those made under the direction of the Land Commission, commonly known as "kuleana" surveys, had the same defects as the first surveys in most new countries. There was a lack of proper supervision, no Bureau of Surveying, and the president of the Land Commission was overwhelmed with work such as there was no time for supervision. In addition, there was little money to pay for the work and little time to accomplish it.

No uniform rules or instructions were given to the surveyors, who were "practically irresponsible." Few of them were regarded as thoroughly competent surveyors, while some were not only incompetent but "careless and unscrupulous." The result was that almost every possible method of measurement was adopted. Some used 50-foot chains, while others used four pole (66-foot) chains divided into links; some attempted to survey by the true meridian, others by the average magnetic meridian, while most made no allowance for local variations of the needle. Some recorded surveys have been found to have been made with a ship's compass and even a pocket compass. Few made the effort to mark corners or to note topographic features.

In addition, rarely was one section or district assigned to one surveyor. It has been reported that over a dozen were employed in the surveying of Waikiki, for instance, not one of whom knew what the other surveyors had done or made any attempt to make his surveys agree with theirs where they adjoined one another. As one might expect, overlaps and gaps became the rule rather than the exception, so that it became generally impossible to put these old surveys together correctly on paper without determining their true relative locations through actual measurements on the ground.

There are two types of recording systems in the State of Hawaii, the Regular System and the Land Court System, the latter of which is a modification of the Torrens System and includes a Land Court. (See Section 7.3 on juridical cadastres.)

As can be seen, the land grant and tenure systems of what is now the State of Hawaii is a mixture of conveyances and processes over a period of time. Despite the plan to confirm titles and organize lands, the net result is a variety of categories of land tenure and land records. While many surveys were performed, and are available, they often provide not the most desirable definition of claims.

Based on the previous discussion, it may be seen that while the U.S. colonies were influenced by European concepts, mostly because they were settled by European immigrants, other locations bore resemblance to other parts of the world. For example, the tenure system in the Hawaiian Islands closely resembled the feudal systems of the United Kingdom and her colonies. The homestead system of Hawaii and the homestead system of the Western United States and the setting aside of lands for education in Hawaii and the school lots of the English colonies are among other similarities.

4.5.4 Louisiana

Louisiana was settled under several different governments, which resulted in the French longlots; the square leagues of the Spanish; the irregular subdivisions in the Florida parishes, which developed first under British control; then the American rectangular survey, which encompasses about 85% of the state. French Louisiana was administered as a royal colony during the periods 1699 to 1717 and 1731 to 1763.

Both Spain and England treated parts of Louisiana as royal colonies. Permanent settlement came in 1712, with the early French around New Orleans, who were Creoles, then later refugees from French Acadia known as Cajuns, settling to the west of New Orleans and along the Mississippi to the north. Private companies operated Louisiana from 1717 to 1731. As an example, the Colony of the West, later known as the Company of the Indies, was given control of Louisiana in 1717 and made allodial grants, many of which were quite sizeable in extent.

One feature of land division, based on the period as a French colony, that is prevalent in Louisiana is the abundance of long, rectangular lots. Most of these front on rivers, such as the Mississippi. The lots, also known as arpent lots (from the French system of measurement, both for distance and for area), also in some areas loosely known as river lots, are not confined to Louisiana, however, and are found in many eastern locations, particularly in New England states and most of the original colonies. Michigan and Texas also have many of these types of lots, the former having a designated area known as the French Lots.

The longlot system was a natural fit to the levees and bayous of the southern part of Louisiana. Fronting on the stream and river banks and extending inland toward the back swamp land of the floodplain provided an owner with a proportionate share of the best soils near the river and the wetter soils toward the back. In addition, there was access to the water body for fishing and transportation. Marking the longlots was simple, since at first, only the corners at the water were marked and directions given for the sidelines. This system was a very workable one since rivers were often used as routes of travel; therefore, lots were accessible by water, then extended back from the water such distance to provide sufficient acreage for its intended purpose. Contrast this with the *moku-ahupua* system of Hawaii.

Many of the longlots were not surveyed when granted but became fixed when American surveyors laid out townships and sections. The French were resistant to the American surveying rectangles, believing that it made no sense to lay off squares that bore no relation to the usable land and might not have access to the streams. Eventually, the American surveyors were authorized to survey land along streams in longlots of 160 acres.

During Spanish control, which began in 1770, another type of land grant, known as the *sitio*, was introduced in areas west and northwest of the French areas. Those grants were usually square but often not oriented according to the cardinal directions and often a league on a side.

The Florida Parishes of Louisiana contain a mixture of French, British, and Spanish royal grants; irregular American grants made to squatters; and later grants conforming to the official United States survey.

4.5.5 Texas

Texas is included within the states known as the metes and bounds states, although a number of original surveys and conveyances were made based on one or more rectangular grids. About 15% of Texas was originally granted to private interests under Spain and Mexico, with the remaining 85% distributed according to the policies of the Republic, then State, of Texas. None of this large state was ever included as part of the public domain.

European dominion in Texas began in 1716, when Spain established missions that were followed by presidios, farms and grazing settlements, and towns. It is estimated that at least 10 million acres of land had been granted during Spanish control. Texas lent itself to certain aspects of Spanish land division, similar to the Louisiana system of longlots.

The *sitio* grants, discussed earlier in Louisiana, actually were made much earlier, in Texas. The unit of land measure used by the Spaniards was the square league, equivalent to 25 million square *varas* or 4428 acres. A league was the standard *sitio de granado mayor* for the raising of cattle and horses; a half-league (half a league on each side) could also be made. The phrase "one league to each wind" was coined as the surveyor's description of a square four-league tract more or less oriented with the compass.

Longlots, first laid out in 1731 as irrigated plots or *suertes*, were also used in Texas in Spanish layouts, but in a variety of forms. The river lots were used in the nonirrigated land along the lower Rio Grande River and its southern tributaries. In 1767, the policy was to grant original settlers two leagues of grazing land and 12 *caballerias* (400 acres) of cultivable land. More than 300 of the great longlots, known as *porciones*, were laid out along the Rio Grande in the 1760s, the largest of which was 5 miles wide by 17 miles long, containing well over 10 leagues of land.

Moses Austin received a contract to introduce American settlers to Texas and in 1820 presented his colonization plan at San Antonio. It was met with disfavor, but shortly thereafter, Austin met up with Baron de Bastrop, a friend who he knew earlier in Louisiana. De Bastrop was friends with the governor and the plan was approved; however, Austin died soon after and his son took over the project. While the younger Austin was getting organized, Texas passed from Spanish to Mexican sovereignty, causing Austin to obtain new approvals from Mexico City.

The Austin's colonization plans initiated the *empressario* system of land grants, which governed most of the settlement of Texas during the Mexican era. An *empressario* was given authority to grant lands to settlers within an assigned area in return for a piece of land upon settlement of a minimum number of families. Between two and three dozen *empressario* contracts were made, nearly taking up the 150-mile-wide strip of coastal Texas between the Sabine and Nueces rivers. Although most of these did not meet the expectations of settlement, overall, the settlement was successful.

This early Austin plan resembled the early colonial headright scheme with American land measures, although Mexican law required Spanish measures for the *empressario* grants, which were therefore defined by leagues and labors. Adjoining the Austin colony to the southwest was the De Witt Colony. Holdings were as rectangular as was possible fronting on the Guadalupe River or a tributary thereto and extending back at right angles. Both of these colonies set aside four-league reserves fronting on rivers for their principal towns. The towns themselves were mostly either rectangular or square.

Texas became a republic in 1836, formed as a separate nation after gaining independence from Mexico. The Republic of Texas claimed boundaries that included all of the present U.S. state of Texas and also parts of present-day Oklahoma, Kansas, Colorado, Wyoming, and New Mexico based upon the Treaties of Velasco between the newly created Texas Republic and Mexico.

The Republic immediately began offering rewards in land to soldiers who would participate in the military. The first military bounty was in 1835, and it abandoned Mexican units of measurement. This was followed by the headright system of land granting, matching the Mexican standard of a league and a labor. Homesteads and preemption claims were also part of the process, as well as, for a brief time, *empressario* grants.

Texas was admitted to the Union in 1845, although there was no immediate change in their land system. The public lands continued to belong to the State of Texas and at the disposal of its government. Land transfer was by farm-size plots of 80 to 1280 acres, through the sale of scrip, preemption, homesteads, and veteran's bounties and donations. In addition, more than 30 million acres were granted to railroads, along with 50 million acres in support of schools, the university, and other educational endeavors. Railroads were granted 8, 16, or 20 sections per mile as encouragement. They were required to survey the lands into a checkerboard pattern, receiving the odd-numbered squares. The even-numbered squares were turned over to the public school fund, resulting in the surveys automatically divided between the railroads and the schools.

When Texas was annexed into the United States, the State kept control of all of its public lands. As a result, Texas is the only U.S. state to control all of its own public lands, unlike the PLSS states. Today, the head of the General

Land Office (GLO) is the commissioner and is an elected official. The office's main function is to manage the public lands of Texas. When instituted, the GLO was overseen by a surveyor-general, under whose instructions were deputy surveyors. Their task was to lay off assigned portions of territory into sections, ranges, and townships; make maps thereof; and return them to the surveyor-general.

With the variety of granting systems in Texas under several different forms of government, timing is critical along with the history of the formation of the state, in order to understand the origin and development of a tract of land. More complicated than most states, the history of Texas serves as an important example of the importance of the history of a territory and the changes in its laws and policies over time.

4.5.6 Significance of Original Cadastral Surveys

The original creation of title is typically based on the original cadastral survey, which was used to describe land being transferred from a sovereign to a private land owner. Thus, knowledge of the cadastral system and the conveyance under which the title to land is created can be significant to a contemporary title. Not infrequently, in-depth research and investigation into the original conveyance disclose outstanding rights created a century or more ago and never extinguished. There have been several recent examples that are noteworthy:

In the case of *Ski Roundtop v. Wagerman*,* the two parties in the action both claimed a part of each others' land, creating an overlap. It turned out that neither party had record title to the disputed strip. The court referred to a previous Maryland coal case, *Maryland Coal and Realty Co. v. Eckhart*,† which contained a historical account on the law of land patents. This case describes Maryland's authority to patent lands derived from its sovereign heritage. The original charter, from King Charles I in 1632 to Cecil Calvert, gave him exclusive power to grant lands in Maryland, with that power vesting in the State after the American Revolution. The court stated, "Although the case law in this area is sparse, it appears that a requisite for valid title to real property is an original conveyance of public land by the State. See 3 American Law of Property, § 12:16 (1952); 73B C.J.S., Public Lands, § 188 (1983); 2 Patton on Titles, § 281 (2d ed. 1957). Absent such a conveyance, one purporting to transfer an ownership interest in such property transfers nothing, and no quantity of successive transfers by deed nor the mere passage of time will metamorphose good title from void title."

The Massachusetts Land Court decided a case in 2008 concerning a way created in the 18th century. In *Campbell & Others v. Nickerson & Others*,‡ an

* 79 Md.App. 357, 556 A.2d 1144 (Md.App. 1989)
† 25 Md.App. 605, 337 A.2d 150 (Md.App. 1975)
‡ 73 Mass.App.Ct. 20 (2008)

easement was established with a conveyance by deed in 1711. The instrument conveyed shares in a large parcel of land to 14 grantees, known in this case as the "proprietors." In 1714, at a proprietors' meeting, they divided the lots among themselves and stated that they "and their heirs and assigns forever to have and receive…privilege of passing through one or another of said lots…for egress and regress."

The court determined that the rights were to apply to the lots not only as then laid out but also to lots "as they may be [laid out] in each respective division." The judge reasoned that the proprietors intended the rights of passing over the lots to apply to future divisions of the land and that those rights were established in a manner comparable to the later "common scheme" doctrine.

In 1971, the New York court recognized the value of history and its relation to the original creation of title in the case of *Dolphin Lane Associates, Ltd. v. Town of Southampton*.* It stated, "this action, by its very nature, involves the tracing of chains of title going back to the earliest settlement of the Town of Southampton. Thus, the history of the early settlement and development of Southampton must be examined. The past must be explored in order to understand the present." Whereupon the court proceeded to outline in some detail the evolution of the Town of Southampton, from 1639, when eight men from Massachusetts agreed to establish a permanent settlement, through the addition of 10 more known as undertakers, a listing of applicable patents, overlapping claims from the Dutch settlers of New Amsterdam as well as the Colony of Connecticut, and the eventual repatenting and subsequent confirmations thereof.

Ultimately, the title was vested to all land, not in the lawful possession of some individuals, within the boundaries named in the Town of Southampton. The legal title to this land was held by the Trustees of the Freeholders and Commonalty of the Town of Southampton.

The Pennsylvania case of *Bruker v. Burgess and Town Council of Borough of Carlisle*† had to do with a town square having been used for more than 200 years, which the court acknowledged that such use "is sufficient to raise a conclusive presumption of an original grant for the purpose of a public square; such is an ancient and well established principle of law. Nor can it be denied that, where such a dedication has been established and the public has accepted it, there cannot be any diversion of such use from a public to a private purpose, and it is also true that, where a dedication is for a limited or restricted use, any diversion therefrom to some purpose other than the one designated is likewise forbidden." In that case, it was found that, "The town of Carlisle was laid out in 1751 according to a plan of Thomas Penn, then Proprietor of Pennsylvania." In a letter to Richard Peters, Secretary of the Province, he wrote that in the center of Carlisle,

---

* 339 N.Y.S.2d 966, 72 Misc.2d 868 (1971)

† 376 Pa. 330; 102 A.2d 418 (Pa., 1954)

he would have a Square and that "the Court House may be in the middle of one side and the Gaal in any place near, there may be a places I think in the middle of the Center Square for a Market, or if that will take off too much of the lots, a lot may be given at the rent of a Shilling as for the other publick used for that purpose..." A public market occupied a portion of the southeast quarter of the Square as early as 1764, and from time to time thereafter, new market houses were erected on the same site and markets were regularly conducted there.

These modern examples underscore the importance of historical consideration of any title and an evaluation of its origin. An understanding of the evolution of any particular state is paramount in order to fully comprehend many of the aspects of a tract of land. This is true even when title is established under one system but is now within a different system. Many original surveys establishing boundaries were performed in a rectangular system, such as those described in this section, but later converted to a different system, often metes and bounds or a description based on adjoining lands. As a result, a title researcher may be misled as to the true description, and therefore boundaries, of the subject parcel. This is far less common in the public land states in that the majority of titles may be traced back to the sovereign based on the official government survey establishing the beginning source of title. Regardless of the type of cadastral system, tracing a title back to the originating patent results in a complete picture of the chain of title and reflects any changes over time.

The nature of the original cadastral survey also has had a significant impact on the landscape of areas. The original creation of land lines dictated the patterns followed and perpetuated later in the subdivisions of them, which is often visible today on aerial photography and satellite imagery. In many cases, the original lines themselves, including the major grants, may be seen on the imagery. As examples, the early large tracts of land such as those acquired and surveyed for William Bingham in New York, William Penn in Pennsylvania, and the Forbes Purchase in Florida created significant impacts upon the use of the land and its ultimate subdivision and transfer.

As may be seen from the information provided in this chapter, there has been great variation in the cadastral systems used in distribution of land in the United States. The various title concepts and surveying practices associated with those systems, the procedures of the various court systems concerning overlapping titles and other conflicts, as well as the consideration and understanding of the very basis of original titles by Native Americans and/or European powers have left a complex matrix demanding considerable study for boundary survey practitioners.

All this serves to illustrate the importance of original grant history and the establishment of original titles. Many of these grants were surveyed and monumented, creating original lines and corners. Not all have been meticulously maintained over time, nor have they all been diligently sought by retracement surveyors and others. Records in the form of the grants themselves, survey maps, and data, some of which have been lost or misfiled, are often found in obscure places, although a surprising amount is still in existence. Occasionally, something surfaces, often throwing modern surveys and titles into question. An example is the 1607 map of the Popham Colony in Maine. In 1888, the map was discovered in the Spanish archives, purporting to be that of the early colony. Subsequent investigation over a period of more than 100 years proved that it was indeed authentic and of great value and historical significance.

Recommended Additional Reading on Subdivision of Land in the United States

Alexander, William DeWitt (1882). *A Brief History of Land Titles in the Hawaiian Kingdom.* Honolulu: F.C. Advertiser Co. Steam Print.

Cadle, Farris W. (1991). *Georgia Land Surveying History and Law.* Athens: The University of Georgia Press.

Chandler, Alfred N. (1945). *Land Title Origins.* New York: Robert Schalkenbach Foundation.

Cazier, Lola (n.d). *Surveys and Surveyors of the Public Domain 1785–1975.* Washington, DC: U.S. Government Printing Office.

Ford, Amelia C. (1910). *Colonial Precedents of Our National Land System as It Existed in 1800.* Bulletin of the University of Wisconsin, No. 352. Madison: University of Wisconsin.

Hawes, J. H. (1882). *Manual of United States Surveying.* Philadelphia: J.B. Lippincott & Co.

Johnson, Hildegard Binder (1967). *Order Upon the Land.* New York: Oxford University Press.

Pattison, William D. (1964). *Beginnings of the American Rectangular Land Survey System, 1784–1800.* Department of Geography Research Paper No. 50. Chicago: The University of Chicago.

Price, Edward T. (1995). *Dividing the Land.* The University of Chicago Geography Research Paper No. 238. Chicago: The University of Chicago Press.

Thrower, Norman J.W. (1966). *Original Survey and Land Subdivision.* Chicago: Rand McNally & Company.

White, Albert C. (n.d.). *A History of the Rectangular Survey System.* Washington, DC: U.S. Government Printing Office.

5

Land Descriptions

5.1 Types of Boundaries

Central to an understanding of land descriptions is an understanding of the concept of land boundaries. Land boundaries are lines that limit or separate ownership or jurisdiction. Therefore, land boundaries may be lines between adjoining private lands, between private and public lands, between adjoining public jurisdictions, or limits of certain types of regulatory jurisdiction.

There are two key attributes of a good boundary: First, the boundary should be more or less permanent. Second, the boundary should be recognizable and should be locatable without ambiguity. In addition, in earlier times, it was desirable for a boundary to be physically defendable. Examples of easily defended boundaries are rivers, coastlines, and the ridgeline of mountain ranges. In more recent time in developed nations, the advent of a legal property system provides a means of defending boundaries through the legal system. Thus, this has largely eliminated the need for physically defending most boundaries, with the possible exception of national boundaries.

Boundaries may be based on three types of features: natural objects, manmade objects, and legal or mathematical entities. As examples of boundaries based on natural features, a boundary may consist of the thread of a stream, along or offset a given distance from the mean high water line of an ocean, along the edge of a wetland as determined by vegetative indicators, or a line run between two prominent trees or other physical features. As examples of boundaries based on manmade objects, a boundary may consist of a line along a manmade ditch, berm, or roadway or a line between two concrete monuments. As examples of boundaries based on legal or mathematical entities, a boundary may consist of the limits of a quarter section of the U.S. Public Land Survey of a lot in a recorded private subdivision, a line along or offset from the east line of a U.S. Public Land Survey section, along a series of distances and directions starting at a corner of a recorded subdivision, or lines between points defined by geographic or state plane coordinates. Another example is a description based on the center of a U.S. Public Land Survey section. While the center of a section is not a physical government monument, it is a point capable of mathematical ascertainment, "thus constituting it, in a legal sense, a monument call of the description."* Similarly, a called-for adjoiner, or adjoining tract, may be classed as a monument,† provided it is called for, existing at the time of the conveyance, clearly established and identified, and accurately located.

Following are two quotes from the old testament of the *Holy Bible* illustrating early property descriptions using natural features for boundaries and the use of manmade monuments (rock cairns) for inflection points along boundary lines.

> And as for the western border, ye shall even have the Great Sea for a border; this shall be your west border. (Numbers 34:6, *The New English Bible*)
> So Jacob took a stone and set it up as a pillar. He said to his relatives, "Gather some stones." So they took stones and piled them in a heap.... Laban said to Jacob, "Here is this heap, and here is the pillar I have set up between you and me. This heap is a witness, and this pillar is a witness, that I will not go past this heap to your side to harm you and that you will not go past this heap and pillar to harm me." (Genesis 31:45–52, *The New English Bible*)

5.2 Purposes of Descriptions

Throughout history, mankind has created representative systems, such as written music notation and currency, that have allowed great advances in civilization. One such system, the representation of land with a written description, is especially important to property law. The fact that a parcel of land can be represented by a written document that uniquely identifies the parcel allows for the land to be the subject of legal transactions. In addition, the written description allows for the land to be recorded in a central deed repository and also located on cadastral maps. It allows the described land to have a parallel life as a capital asset in addition to its physical life and ability to support human life. Thus, descriptions have played an important part in aiding the protection of claims to land under legal systems, as well as allowing the consideration of land as a commodity.

5.3 Types of Descriptions

Descriptions referencing formal (recorded) subdivisions—A very common, and simple, type of description is that referencing formal subdivisions of land.

* *Matthews v. Parker*, 299 P. 354, 163 Wash. 10 (1931)

† A line called for is quite as controlling as any natural or artificial boundary. *Parran v. Wilson*, 154 A. 449, 160 Md. 604 (1931); *Ramsay v. Butler, Purdum & Co.*, 129 A. 650, 180 Md. 438 (1925)

This type of description involves a reference to a lot or lots within a subdivision of land where a plat of survey has been recorded in the public records.* Descriptions of this type may involve a reference to subdivisions by government agencies such as those performed as a part of the Public Land Survey System (PLSS) or those created by private landowners and recorded in the public records. Such recorded subdivision plats are typically filed in each county courthouse. Most states have statutes providing standards for the survey and platting of such subdivisions. In addition, many counties have subdivision ordinances with additional requirements. Therefore, modern-day subdivision plats are relatively standardized. Since such subdivision plats are available in the public records, a description of land portrayed on such a plat requires only that the lot or lots be identified along with the subdivision name and a reference to where the plat is recorded. With such a description, a plat referenced in a description becomes a part of the description, "with as much effect as if copied into it."†

> Example: Lots 100 and 101 of Dreblow & Company's the Silver Lake Subdivision as recorded at Plat Book A, Page 3 of the public records of Jefferson County, Florida.

One of the most common problems with subdivision plats is that, frequently, one will not agree 100% in all aspects with the evidence set by the original surveyor on the ground. When encountered by some surveyors, the tendency is to set an additional marker to conform with the distance(s) stated on the plat, sometimes subsequently incorporated in the resulting deed description. Several recent judicial decisions and at least one older decision state that this is an incorrect approach, and when such a disagreement is found, it is to be resolved against the plat.‡

Metes and bounds descriptions—In those U.S. states not subdivided under the PLSS as well as in the U.S. territories, many of the surveys for the original private conveyances were made by the metes and bounds method. Metes and bounds descriptions are also used extensively to describe irregular areas being severed from lands originally subdivided under the PLSS. Examples include mineral surveys, homestead entry surveys, donation tracts, and townsites.

---

* Instances exist where subdivisions have been done and conveyances made from them that have never been recorded. Many courts have upheld such practice. Modern-day subdivisions result in a plat, but earlier subdivisions have been in the form of sketches, and a few that contained numbered lots from a scheme in the developer's head, without anything having been drawn on paper. See *Perkins v. Jacobs*, 129 A. 4, 124 Me. 347 (1925).

† *Jacobs v. All Persons, etc.*, 106 P. 896, 12 C.A. 163 (1910); *Goldsmith v. Means*, 158 A. 596, 104 Pa. Super. 571 (1932)

‡ *Tyson v. Edwards*, 433 So.2d 549, 552, Fla. 5th DCA (1983). *See Outlaw v. Gulf Oil Corp.*, 137 S.W.2d 787, Tex.Civ.App. (1940) and *Bean v. Batchelder*, 78 Me. 184, 3 A. 279 (1886).

A metes description identifies a beginning point and then describes each course in sequence around the perimeter of the tract until the point of beginning is reached again to complete the description of the perimeter. Such a description includes a distance and direction for each course. A bounds description defines the boundaries of a tract of land by identifying adjoining properties but does not typically provide a direction. A metes and bounds description provides measurements as well as adjoining properties, where practical, as well as calling monuments.

Examples:
Bounds description: All of that land lying north of State Road 99; bounded on the north by land of Albert Bowie, bounded on the east by lands of Betty Anderson, and bounded on the west by Trout Creek.

Metes description: A tract of land in Section 12, Township 3 North, Range 6 East in Jefferson County, Florida, more particularly described as follows: For a point of beginning, commence at an old axle marking the southeast corner of said Section 12; then go N00°01′E for 200 feet; then go N89°59′W for 400 feet; then go S00°01′W for 200 feet; then go S89°59′E for 400 feet to the point of beginning.

Metes and *bounds description*: A tract of land in Section 12, Township 3 North, Range 6 East in Jefferson County, Florida, more particularly described as follows: For a point of beginning, commence at an old axle marking the southeast corner of said Section 12; then go N00°01′E for 200 feet along the east line of said Section 12 to an iron pipe; then go N89°59′W for 400 feet to an iron pipe; then go S00°01′W for 200 feet to an iron pipe; then go S89°59′E for 400 feet along the south line of said Section 12 to the point of beginning.

As may be seen from the provided examples, metes and bounds descriptions are somewhat more complex than descriptions merely referencing a recorded subdivision plat. Since the survey creating a metes or bounds description is not typically part of the official records, a metes or bounds description must, in effect, describe the results of the survey creating the parcel and not just reference a plat of survey available in the public records.

Other types of descriptions—There are a number of other types of descriptions used for conveying land. Typical of these is a portion of a given tract based on a certain distance or width, based on a certain area, based on a certain fraction of the total area, or based on a manmade or natural feature such as a road or stream. Another common type is a strip description. This type is frequently used for describing a road easement or right of way. Another type of description being increasingly used is based on geographic coordinates. Many descriptions also use a combination of the previously described types.

Examples:
Division by width: the westerly 50 feet of Lot 2…
Division by area: the southern 10 acres of Government Lot 2…
Division by fraction: the western one-half of Lot 4…

Division by feature: all of Section 12…lying northerly of U.S. Highway 90.

Strip description: A right of way for ingress and egress purposes across a strip of land lying 30 feet on each side of the following described center line…

Description by coordinates: …For a point of beginning, commence at an old concrete monument marking the south quarter corner of said Section 24 and having a north coordinate of 1,972,048.50 and an east coordinate of 563,589.10; then go N00°01′E for 200 feet to a point having a north coordinate of 1,972,248.62 and an east coordinate of 563,589.16; then go N89°59′W for 400 feet to a point having a north coordinate of 1,972,248.5 and an east coordinate of 563,589.16; then go S00°01′W for 200 feet to a point having a north coordinate of 1,972,048.50 and an east coordinate of 563,589.10; then go S89°59′E for 400 feet to the point of beginning. Coordinates are in feet and are based on the Florida State Plane Coordinate System, North Zone, NAD88.

PLSS descriptions—Descriptions based on the PLSS have traditionally been some of the most commonly used in the United States in the past. Thirty states were subdivided using the rectangular system, as described in the previous section. Therefore, almost all of the privately owned lands in those states were originally conveyed by reference to that system.

The original 1796 Public Land Act specified that public land was to be sold by the section. Since each square-mile section, with a standard size of 640 acres, had its own identity, conveyance of any of the public lands that had been surveyed under the PLSS involved a relatively simple description providing the pertinent section, township, and range. In addition, the description needed a reference to the base line and meridian from which the section had been surveyed and to the pertinent official plat that depicted the section.

Example: Section 6, Township 1 North, Range 2 East, Tallahassee Baseline as depicted on the official General Land Office plat approved June 1, 1842.

Due to concern that the minimum size of 640 acres for land purchases restricted land ownership to the wealthy, successive legislation reduced the minimum size for purchase of public land. In 1800, the half section, with a standard size of 320 acres, became the minimum size. In 1804, the sale of land by the quarter section, with a standard size of 160 acres, was authorized; and in 1820, sale of land by the half quarter section, with a standard size of 80 acres, was approved. Then in 1832, the minimum area for public land sale became the quarter of a quarter section, with a standard size of 40 acres. That size tract became the "modular unit of settlement" and "40 acres and a mule" became the standard requirements considered necessary for subsistence for the average family.

The public land surveys only traversed the perimeter of sections. Thus, all of the boundaries of aliquot parts of sections as described previously were not surveyed in the public land surveys. Nevertheless, the description of such tracts is still a simple process. Figure 5.1 illustrates the convention used. Note that when attempting to visualize the location of aliquot parts of sections, it is best to start at the end of the description and work backward.

> Example: The northeast one quarter of the northeast one quarter of Section 11, Township 1 North, Range 2 East, Tallahassee Meridian as depicted on the official Public Land Office plat approved June 1, 1842.

In the platting process for the public lands surveys, it was the practice to divide fractional sections along navigable water bodies into relatively small lots. Typically, the boundaries of such lots were drawn along lines of fractional parts of the section. This was a cartographic process, and the boundaries of such government lots were not surveyed, except for any boundaries coincident with section lines and those coincident with meander lines along the water body. As with descriptions for PLSS sections and aliquot parts of sections, government lots may be described simply by reference to the lot number, section, township, range, base meridian, and pertinent official plat.

> Example: Government Lot 3, Section 12, Township 1 North, Range 2 East, Tallahassee Meridian as depicted on the official General Land Office plat approved June 1, 1842.

Descriptions according to lot and range—Throughout the so-called nonrectangular states and areas (non-PLSS) are found individual, stand-alone,

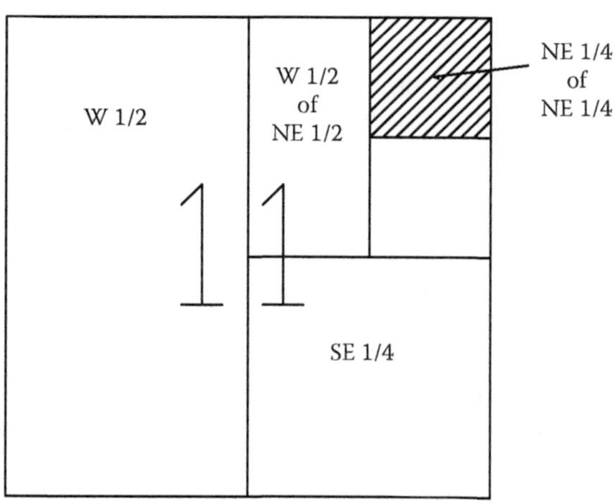

FIGURE 5.1
Fractional parts of public land survey sections.

rectangular divisions (see Chapter 4 for a detailed discussion). Typically, townships were divided into lots of a specified size and arranged in tiers or rows, called ranges. While considerable variation exists in size and shape, many were rectangular in shape, measuring 100 rods by 160 rods. This was a convenient combination to work with, and to divide, since any number multiplied by 160 rods equals that (same) number of acres. As a result, the 100 × 160 rod lot was 100 acres in size, being an ideal size for the average farm.

One of the modern problems with this type of system is the manner in which the original lots were laid out by survey. The original surveyors, realizing that measuring errors occur, would insert what was known as "allowances," some of which were standardized, e.g., "Common Allowance of about Twelve Chains to a mile for Sagg of Chain." This practice led to overrun in acreage in order to ensure that original lots were not too small, thereby shortchanging the recipient. Later retracement surveys frequently did not take such allowances into consideration, resulting in discrepancies within and between subdivided lots.

Where lots overrun in area (acreage), special care must be taken with the interpretation of description. For example, considering that a 100-acre lot may overrun considerably in total area, a 110-acre lot (actual amount, but reported to be 100 acres) could be divided in two basic ways: by size or by area. Conveying *one-half* of the lot, purported to be 100 acres, would result in a 55-acre lot (1/2), although reported to be 50 acres. Conveying a specified acreage, such as 50 acres, would result in a 50-acre lot, as reported. However, the remainder, also reported to be 50 acres (thought to be 1/2), would actually contain 60 acres (110 − 50 = 60).

Descriptions based on the longlot system—Also known as arpent lots and river lots, this French system was used extensively along water bodies (lakes and rivers) in some areas. Common in Louisiana and Texas, it also appears in areas settled by the French throughout the eastern states, Maine, New Hampshire, and Michigan being among the most common. Basically rectangular, or nearly so, the lots front on the water, providing access and, depending on the width, extend approximately at right angles back into the interior a specified distance to contain the desired, or planned, area. Generally, these lots were surveyed and numbered according to a specified plan.

Descriptions based on the Rancho System—Ranchos were generally large tracts of land granted by the Spanish or the Mexican government. They were of the metes and bounds type and often consisted of thousands of acres. Even though they are mostly found in the so-called rectangular (PLSS) states, they are independent of that system and were subdivided as the owner chose rather than by public lands procedures.

Description on one side of a described line—Occasionally, a tract of land is described as being on one side of a line described by directions, distances, and monuments. A description of the original tract must be obtained and the line description applied to it in order to define the parcel being separated.

5.4 Writing Descriptions

The preparation of a land description is an important responsibility. The description must be sufficient to represent the land in various legal transactions, such as buying and selling, serving as collateral for loans, and appropriate taxation, and must also allow for the precise location of the limits of a parcel of land for many years after it is written. Thus, it is critical that the description correctly describe the land, relating it to some well-established and readily identifiable point on the earth (or to a document of record that describes such a point) that will allow the precise location of the parcel on the ground (Robillard and Wilson 2014).

Descriptions are typically written with two major elements, a caption and the body of the description. The caption provides the general location and other identifying information, while the body provides a detailed description of the parcel. In addition, the description may include a qualifying clause that takes away rights to land described in the body and an augmenting clause that adds something to the area described in the body.

> Example:
> (Caption) A tract of land in Section 12, Township 3 North, Range 6 East in Jefferson County, Florida, more particularly described as follows:
>
> (Body) For a point of beginning, commence at an old axle marking the southeast corner of said Section 12; then go N00°01′E for 200 feet along the east line of said Section 12 to an iron pipe; then go N89°59′W for 400 feet to an iron pipe; then go S00°01′W for 200 feet to an iron pipe; then go S89°59′E for 400 feet along the south line of said Section 12 to the point of beginning.
>
> (Augmenting clause) Including an easement for ingress/egress purposes over a 20 foot wide strip running the length of and lying immediately southerly of the described parcel.
>
> (Qualifying clause) The described land is subject to an easement for public ingress/egress over the southerly 20 feet.

The objectives of a description may be summarized as follows:

- To allow for the precise location of the described parcel by a competent surveyor*
- To clearly identify a unique area without ambiguity†
- To allow for determining the relationship of the described parcel to adjoining land
- To be physically and mathematically accurate

* The definition of a legally sufficient real property description is one that can be located on the ground by a registered land surveyor. *Rivers v. Lozeau*, 539 So.2d 1147, Fla. App. 5 Dist. (1989)

† Ambiguity means doubtfulness; doubleness of meaning; indistinctness or certainty of meaning of an expression used in a written description (*Black's Law Dictionary*).

In addition to these objectives, there is an increasing demand for an additional objective, that of providing a link between the described land and its map representation. With increasing reliance on digital maps for various aspects of land tenure as well as many other aspects of our modern world, there is an increasing demand for the inclusion of geographic coordinates in land descriptions for key points on the described tract, as well as associated metadata for such coordinates.

Some important considerations in writing land descriptions are as follows:

- The caption should identify the state and municipality in which the parcel is located.
- Where recorded subdivision plats or other recorded documents are referred to in the description, sufficient information should be provided to allow location of that document.
- One or more ties to monuments of public record should be provided to allow identification and location of the parcel.
- Any monuments called, especially the reference points for metes and/or bounds descriptions, should be permanent in character, visible, and stable.
- The description should not be in conflict with descriptions of adjoining lands.
- The basis for any directions should be stated. For example, bearings should be identified as being based on astronomic, state plane coordinate grids, magnetic north, or as based on a specifically identified line.
- If coordinates of corners are called, the datum for the coordinates should be stated as well as the epoch for the datum.
- The description should contain no ambiguous terms.
- For any curves called in the description, sufficient information must be provided to allow the determination of the beginning and ending points for the curve as well as the radius point. If the curve is tangent (where the radial line at the beginning of the curve is at right angles to the preceding straight line), it is sufficient to provide the arc length, the radius of the curve, and the direction of the curve. If the curve is not radial, then additional elements are necessary. At least two elements of the curve should be stated. The most frequently used elements are the radius, arc length, central angle, and tangent length. In addition, the relationship of the curve to the previous line, the direction of the curve (e.g., concave to the south), and the direction of travel along the curve (e.g., easterly) should be stated.

Example: Then go 1000 feet along a tangent curve to the left, said curve being concave to the north and a radius of 2000 feet.

5.5 Description Plats

Although the written land description is critical to the land tenure process, an accompanying drawing of the described lands is also important to interpretation of the description and its intent. Further, in some jurisdictions, it is a required part of new descriptions. A detailed plot of a description is especially helpful to users of the description, such as land owners, cadastral mappers, land planners, attorneys, realtors, and well as other nonsurveyor users. In addition, it is helpful to other surveyors attempting to survey the described land.

Land description plats are typically prepared on letter-sized sheets for filing in cadastral systems with the written descriptions. For large tracts and parcels with complex boundaries, it is important that the plats be at sufficient scale for viewing. This may be accomplished with the use of multiple sheets or detail enlargements.

An important aspect of a land description plat is the visual hierarchy of the depicted elements to help the viewer focus on the most critical elements. This can be accomplished by the careful use of line weights and pattern types in the drafting process. As an example, the parcel boundary should have the highest priority, while other lines such as the parent parcel boundaries and commencement tie lines should have a lower priority. That difference could be depicted by a heavier weight line for the parcel boundary.

The important elements of an effective land description plat are as follows (Blake 2014):

- A clear depiction of the boundaries of the subject land, along with associated measurements for each segment of the boundary
- Identification of the point of beginning and point of reference
- Identification of the basis for bearings
- Physical address, recordation parcel identification, and other ownership information related to the parcel
- Identification of and relationship to surrounding parcels

Recommended Additional Reading on Land Descriptions

Cuomo, Paul A. and Roy Minnick (1993). *Advanced Land Descriptions*. Rancho Cordova: Landmark Enterprises. With contributions by Michael J. Pallamary.

Robillard, Walter G. and Donald A. Wilson (2014). *Brown's Boundary Control and Legal Principles*. 7th Edition. New York: John Wiley & Sons.

Wattles, Gurdon H. (1976). *Writing Legal Descriptions.* Privately published by the author.

Wattles, William C. (1974). *Land Survey Descriptions.* Several editions privately published by the author; 10th edition revised and published by Gurdon H. Wattles.

Wilson, Donald A. (2015). *Interpreting Land Records.* 2nd Edition. Hoboken: John Wiley & Sons, Inc.

6

Land Boundary Surveying

6.1 History of Boundary Surveying

As discussed in the section on the early history of land tenure, delineation of boundaries is closely linked to the development of property rights. As a result, the need for a delineation process arose as soon as any group or individual claimed an exclusive right to a specific area (Larsson 1991). This is believed to have been associated with more intensive use of land, such as agriculture as opposed to hunting and gathering. Such more intensive land uses required the delimitation of claimed areas to protect the investment in the use of the land.

The process of delineation (or surveying) boundaries may be defined as the establishment or relocation of boundaries of real property. Generally, such surveys are conducted for one or more of the following purposes (Anderson and Mikhail 1998):

- To determine the area and the boundary lines of a tract of land defined by visible landmarks
- To obtain information regarding the boundaries of a tract of land for the purpose of preparing a land description
- To reestablish the boundaries of a tract of land for which a description based on a previous survey exists
- To subdivide a tract of land into two or more smaller tracts

As may be seen from these purposes, boundary surveying requires a marriage of technical skills and legal principles. Considering these purposes, there has obviously been a need for some aspects of boundary surveying as long as there have been individual claims to tracts of land. The earliest documented accounts of boundary surveying have been found in Egypt, Babylon, and in Mesopotamia in current-day Iraq. As examples, clay tablets depicting land surveys have been found in Mesopotamia that date back to about 2400 BC. Engravings illustrating an early Egyptian survey crew surveying boundaries along the Nile River have been found in a tomb erected between the 16th and 17th century BC in Thebes, and a boundary stone set about 1200 BC and inscribed with the name of the surveyor has been found in Babylon. Other mentions of boundary surveying may be found in ancient literature, including Homer's *Iliad,* which was written about the 9th or 8th century BC, and the *Holy Bible.*

The practice of boundary surveying has been held in high esteem throughout early history, and surveying was considered one of the liberal arts. The ancient Greeks laid out their towns using rectangular surveys and that practice was later adopted in Rome. During that era, surveyors also performed tasks such as marking boundaries and mediating boundary disputes. During that era, the most important instruments used were the *groma,* or surveyor's cross, which was used for measuring angles or straight lines, and the *decempeda,* a 10-foot-long measuring pole. Imported stones, engraved with pertinent information, were used for boundary markers so that they could be readily identified (Glick et al. 2005).

In a treatise on the development of mathematics in the early middle ages, the practice of estate division was used as an example, with comments such as the following (Glick 2005).

> Simplified Roman procedures appear to have been used by individual farmers while triangulation was associated with institutions that commanded the services of professional surveyors (such as the Monastery of Ripoll which was acquiring huge donations of land during the tenth century). In the course of repartimiento, parcel boundaries were simply measured with a cord (soga) along a straight boundary such as a road or canal (the basic Roman agrimensorial procedure). But when the boundary was not clear, triangulation was resorted to. Large-scale or difficult surveys were carried out by specialists in land measurement (the muhandis in Al-Andalus, the soguejador in eastern Spain) and in the surveying, more specialized, of irrigation canals.

The art of surveying declined considerably during the early Middle Ages. However, it was revived by the late Middle Ages when, for fiscal reasons, the measurement of land became more common again. Literate people began to specialize in surveying, and practitioners began to write descriptions of their activities. Such writings described the marking of property boundaries, placing boundary stones, and calculating areas of property (Glick et al. 2005).

In the New World, boundary surveying took on an especially important role. Early on, there were no maps to rely upon, and maps were made by explorers and others as they traversed the coastal areas and sailed inland as far as they could up rivers flowing into the ocean. Consequently, the first maps were limited, delimiting only what travelers could actually see. After that, as explorations and settlements proceeded inland from the coast, more sophisticated maps were developed, and larger areas depicted. These maps later became the guidepost for future exploration and settlement.

The American continents contained vast areas, relatively undeveloped by European standards when colonization began. Thus, unprecedented challenges were involved in each colony in defining, marking, and recording each settler's land. It was not simply a matter of locating existing land claims. The large influx of new settlers resulted in a significant demand for men with the knowledge of mathematics and astronomy and skills necessary to survey the American continents. Although the surveyors in the American colonies came from a wide variety of backgrounds, they usually held high social positions and did well financially. As a result, many of the early American surveyors had prominent roles in the early development of the United States.

One especially prominent example is George Washington, the first president of the United States. Washington studied and was fascinated by surveying as a young boy and would often impress visitors to his boyhood home by demonstrating how he could survey and prepare a resulting plat for his brother's turnip field. As a youth, Washington was an ideal field surveyor. At the age of 16, he was a strapping young man of six feet in height with a size 13 shoe and having hands described as the largest that "I have ever seen on a human being." At that age, he was hired as an assistant surveyor to survey a 5,400,000-acre land grant in the Shenandoah Valley by Lord Fairfax. In the same year, he was recommended by Lord Fairfax and subsequently hired as the official surveyor for Culpepper County, Virginia. He was active as a surveyor as a young man, and the public records in Virginia contain hundreds of his survey plats (Woodward 1926).

Another prominent example is the third U.S. president, Thomas Jefferson. Jefferson came from a long line of surveyors, including his father, grandfather, and great grandfather. When Jefferson was 14, his father died and bequeathed him surveying books and instruments in addition to the largest portion of his estate. Although Jefferson studied the law and began his career as an attorney, he soon became disenchanted with a legal career. As a result, in 1774, he gave up his law practice and was appointed as surveyor of Albemarle County, Virginia, a position that his father had held before him (Schachner 1951). During Jefferson's presidency, his accomplishments included several of great importance to modern-day surveyors. He established the Survey of the Coast as the nation's first scientific agency. That agency later became the U.S. Coast and Geodetic Survey, currently a component of the National Oceanic and Atmospheric Administration (NOAA), and is responsible for the geodetic framework of the United States as well as the charting of the nation's coastal waters. Another of Jefferson's accomplishments significant to modern surveying was laying the groundwork for the U.S. Public Land Survey System and standardizing a system of measures.

Two other prominent gentlemen from the early days of the United States also spent a portion of their careers in the surveying profession. These are the 16th president of the United States, Abraham Lincoln, and philosopher Henry David Thoreau.

6.2 Land Boundary Delimitation

Boundary surveying may be classified into two distinct categories—original surveys and retracement surveys. Original surveys are typically subdivisions of a parcel of land into two or more smaller tracts to create new parcels that may be uniquely identified by descriptions and/or plats. Such original surveys include those performed as part of the U.S. Public Land Survey System, surveys of a larger parcel to cut out a new parcel, or more formal subdivisions of large tracts into residential lots. Retracement surveys involve a survey of a land parcel that has been previously surveyed and for which either a graphic or written description exists based on the original survey.

6.2.1 General Procedures

Requirements for original boundary surveys are generally prescribed by governmental regulations in today's society. For example, procedures for the U.S. Public Land Surveys are detailed in the *Manual of Surveying Instruction* (Bureau of Land Management 2009). Similarly, most municipalities have subdivision ordinances that prescribe requirements for formal subdivisions. Procedures for retracement boundary surveys, the emphasis of this section, are generally not prescribed by regulation. Rather, the original location of the boundary must be found by application of the art of surveying using appropriate legal precedent. The aim of a retracement survey is to follow, as closely as possible, in the footsteps of the original surveyor and to reestablish boundaries in the exact position in which they were originally placed. Therefore, the goal of the field work for retracement surveys is to find evidence of the original position. When a retracement survey fails to find satisfactory evidence of the exact original location, other measures, such as apportionment of measurements, may be used to establish the most probable location.

Retracement surveys are often called resurveys in court decisions and in subsequent writings. Nevertheless, a distinction should be made between retracement surveys and resurveys, since they each have a special function, and each of the terms connotes different procedures. Strictly speaking, a resurvey is a correction survey of an original survey because of one or more irregularities in the latter. Resurveys "correct" or "fix" an original survey with a new survey. They cannot be done unilaterally by a surveyor even if gross errors are encountered in a survey subsequent to the original survey, nor may they be corrected by the court system since they do not have the authority to do so. Only an Act of the U.S. Congress may sanction a resurvey of a federal survey. Private surveys demand a different procedure. In either case, land titles that have been created from an original survey cannot be altered without the involvement of the title holders.*

* See *Cragin v. Powell*, 128 U.S. 691, La. (1888).

A retracement survey is the process whereby existing boundaries are tracked and located based on existing evidence. The original survey established the boundary lines, subsequent to which the original (beginning) title was created. Once created, a boundary remains fixed in location from that point on. While ownerships may diminish and expand through sales and acquisitions, once a parcel of land, with its defining boundaries, is created, and it remains fixed forever (Griffin 1960).*

The function and duty of the land surveyor in the surveying and location of lands are best expressed as follows†:

> FIRST, the surveyor can, in the first instance, lay out or establish boundary lines within an original division of a tract of land which has theretofore existed as one unit or parcel. In performing this function, he is known as the "original surveyor" and when his survey results in a property description used by the owner to transfer title to property‡ that survey has a certain special authority in that the monuments set by the original surveyor on the ground, control over discrepancies within the total parcel description and; more importantly, control over all subsequent surveys attempting to locate the same line.
>
> SECOND, a surveyor can be retained to locate on the ground a boundary line which has theretofore been established. When he does this, he "traces the footsteps" of the "original surveyor" in locating existing boundaries. Correctly stated, this is a "retracement" survey, not a resurvey, and in performing this function, the second and each succeeding surveyor is a "following" or "tracing" surveyor, and his sole duty, function and power is to locate on the ground the boundaries corners and boundary line or lines established by the original survey; he cannot establish a new corner or new line terminal point, nor may he correct errors of the original surveyor. The following surveyor, rather than being the creator of the boundary line, is only its discoverer and is only that when he correctly locates it.

In addition to the *BLM Manual of Surveying Instruction*, the federal guide for recovery and retracement of federal lands is *Restoration of Lost or Obliterated Corners & Subdivision of Sections*, a guide for surveyors.§ While these two publications have been generated under the guidance of the Bureau of Land Management for use in surveying federal lands, the general rules are equally applicable to private surveys of public land surveyed areas.¶

---

* Retracement is the process of uncovering physical evidence of monuments and corners by intelligent search on the ground for the calls of the description and field notes of the original survey, guided by the controlling influence of known points.

† *Rivers v. Lozeau*, 539 So. 2d 1147 (1989)

‡ This is a most important qualification.

§ U.S. Department of the Interior Bureau of Land Management, 1974

¶ *Staub v. Hampton*, 117 Tenn. 706, 101 S.W. 776, 781 (1907)

Although procedures for performing a retracement boundary survey will vary considerably with the type of description and other circumstances, the procedures generally follow several basic steps. These are as follows:

1. Data collection—This step, often referred to as "courthouse research" in the past, is possibly the most important part of a boundary survey. This involves the collection of all necessary data for the planning and performance of the survey. It is necessary to have a thorough understanding of the deed for the property being surveyed as well as to obtain any surrounding parcels to correctly perform a boundary survey. As discussed in the previous section, the boundary of senior conveyances of adjoining property may represent the correct boundary line of the property being surveyed. Therefore, the deeds to adjoining property may be as important to a boundary survey as is the deed to the property being surveyed. In addition to deeds, other data that may be necessary include copies of any pertinent subdivision plats, copies of previous surveys, copies of the plat and field notes for any pertinent Public Land Surveys, assessor' plats, and aerial photography, all considered as forms of evidence.

   For some surveys, it is also important to research the entire chain of title back to the original conveyance from the sovereign. (See Section 7.2 for a discussion of this process.) Such a search allows identification of junior–senior rights, as well as the determination of the basis for any bearings used in the description to allow more precise retracement of any existing easements that were not mentioned in the current deed, of any existing boundary agreements that may affect the survey, of any transcription errors that may have been passed down the chain, and of the true intent of the original parties when the subject tract was created (Robillard and Wilson 2011).
2. Data analysis—Once the pertinent data are gathered, an analysis of those data is necessary. Typically, this involves plotting of all of the deeds to scale to allow comparison of the record boundaries of the parcel to be surveyed with surrounding with those of the surrounding parcels. This step should also involve noting the expected location of controlling monuments based on the deeds and examination of aerial photography for evidence of land lines, improvements, and adverse occupation.
3. Survey of visible evidence—This step usually consists of a survey around the perimeter of the property and of any controlling monuments. During that survey, all improvements on or near the subject property and any evidence of easements or encroachments should also be mapped. Ideally, this traverse should be run along the boundary lines themselves to allow better viewing of the relationship of improvements on the subject property and adjoining property to the

line and of any encroachments. Often, however, this is not practical due to vegetation and other obstructions to the line of sight, and a random traverse paralleling the actual boundary lines is necessary.

4. Calculation and analysis of survey results—The survey measurements from the previous step should then be calculated and the results mapped for review and comparison with the deed descriptions. This is possibly the most critical step since it is when preliminary decisions as to the most probable location of boundary corners should be made. Based on the interpretation of the deeds, coordinates for the location of any boundary corners with missing monuments will be made in this step.
5. Search for missing corners and monumentation as needed—Using the calculated coordinates for any boundary corners where monuments were not found in the initial field work, final searches should be made. During this effort, any additional information regarding improvements or other evidence based on the analysis in the previous step should also be located.
6. Demarcation of boundaries—After a thorough search, the location of any corners still missing monuments, as determined by the analysis, should be monumented. An important component of boundary surveying is the demarcation of land boundaries for land management purposes. Other than the survey plat, monuments are often the only visible evidence of the work of the surveyor and are also essential to good land management. Thus, precisely positioned and easily identified monuments are an important product of the surveying. With the ever-increasing value of land resources and associated infrastructure, it is especially important that boundaries be well demarcated with durable, recognizable, and appropriately labeled monuments.

   With the growing requirement for georeferenced boundary surveys, determining precise geographic coordinates for boundaries is becoming an essential part of the demarcation process. As a result, in addition to the traditional boundary surveying process as discussed in previous sections, that process also requires knowledge of more precise surveying methods and a sound understanding of modern geodetic science and coordinate systems.

7. Identification of other evidence affecting land rights and use—A crucial part of the boundary surveying process is the identification of physical evidence affecting unencumbered rights to the land. These include conflicts between descriptions and physical evidence, encroachments, discrepancy between possession and descriptions, the current location of water boundaries or right of ways, evidence of possible prescriptive easements and adverse possession, and access

problems. Problems with land rights due to such problems cannot be discovered by examination of the pertinent documents, but only through field surveys. Other types of evidence affecting land title that can be discovered only through field surveys include changes due to erosion, accretion, and avulsive activity.

In addition, the trend toward increased regulation of land use often requires the location and consideration of various setbacks and regulatory lines such as those for wetlands when conducting land boundary surveys. Although such evidence does not necessarily affect land ownership, it can have a significant effect on permissible land use and land management. Therefore, such information is becoming increasingly important for land tenure as well as for use in cadastral systems, especially multipurpose systems.

8. Preparation of plat and survey results—The final step in a boundary survey is to prepare a plat and report depicting the findings and conclusions of the survey. To ensure usefulness of the survey plat, it should be detailed and complete. To ensure this, it is recommended that plats be prepared in accordance with standards for title surveys developed jointly by the American Land Title Association and the American Congress on Surveying and Mapping. A condensation of those standards follows:
    - The plat should bear the name, contact information, signature, and seal of the surveyor performing the survey, along with the date of completion.
    - The scale of the drawing should be indicated and a graphic scale included. The drawing should be oriented, when practical, so that north is at the top of the drawing and should have a north arrow.
    - The plat should contain all data necessary to mathematically define each line and curve. Both the point of beginning and point of reference for the description, if different, should be shown. Bearings should read in a clockwise direction where practical, and the basis for bearing clearly defined.
    - Both measured and record distances and bearings (or angles) should be shown if they differ. If the record description fails to form a closed figure, that fact should be indicated.
    - In urban areas, record and measured distances to the nearest street right-of-way lines and to the nearest intersecting street should be shown, and evidence of access or lack of evidence to those streets should be indicated.
    - The name, width of right of way, width, and location of pavement relative to the nearest boundary of the tract being surveyed should be shown for all streets and highways abutting the tract.

- Titles and recording data for all recorded plats or similar documents represented by the survey should be shown along with pertinent lot, block, and section numbers. For nonplatted adjoining land, recording data and ownership should be shown. Plats should also show all setback or building restriction lines as well as any gores or overlaps.
- All monuments found or set should be noted, including those beyond the limits of the lands being surveyed.
- All evidence of possession will be shown.
- The locations of any buildings on the land being surveyed will be shown with the perpendicular distance to the boundaries.
- All easements of record and evidence of easements, such as roads, water courses, telephone and electric lines, etc., will be shown.
- The character and location of all walls, fences, and other improvements within five feet on either side of the boundaries will be located and shown.
- Driveways and alleys on or crossing the property will be shown.
- Any cemeteries and burial grounds will be shown.
- Any ponds, lakes, streams, or rivers bordering on or running through the property will be shown. When the property contains a water boundary, that boundary will be located by appropriate methods. The date of that determination will be noted as well as a statement that the boundary is subject to change with time due to natural causes and may or may not represent the actual location of the limit of title.

6.2.2 Evaluation of Evidence in Retracement Surveys

The gathering and consideration of all such physical evidence together with the records from the legal research is a role that has sometimes been called "the art of land surveying." It requires skills in legal research and field measurements as well as a thorough understanding of boundary law to allow the weighing of all of such evidence and make appropriate decisions as to the appropriate boundary location. Today, with our understanding of forensic investigative methods and procedures, evidence recovery and evaluation have been elevated to a unique science. Surveyors are unlikely to deal with DNA, tire tracks, and broken glass. Nevertheless, the reading and interpretation of ancient documents, the study of tree growth (dendrochronology), the classification of types of fence wire, and the identification of corner evidence such as stones and wooden posts have become highly sophisticated. In some instances, they are greatly specialized.

The objective of a retracement survey is to follow in the steps of the survey associated with the original conveyance of the land being surveyed. In doing so, the surveyor must search thoroughly for any evidence of that survey. This should include corner monuments, witnesses established to corners, topographic calls on the original survey plat, blazed trees marking boundary lines, and old fences or structures that may be indicative of boundary lines. Also, the surveyor should not ignore the value of testimony from long-time residents of the area, also known as parol evidence. Such evidence may be of great value, especially where physical evidence cannot be found.

When searching for monuments, the general rule is that a monument must be called in the original description or survey to be legally controlling. Yet, uncalled for monuments may be accepted and held as controlling if they represent the best available evidence of the original location of a corner or have been set by agreement of the appropriate parties.

Searching for wooden monuments or called witness trees often requires spade work. The ground in a suspected corner location should be shaved down carefully to search for decayed parts of the post or discoloration of the soil. Similarly, in searches for witness trees, traces of tree roots or charcoal may often be found at the original location of the tree with careful spade work.

Once all available evidence of the original survey is discovered, that evidence must be evaluated to determine the location of the boundary as originally surveyed. That step is when professional judgment must be exercised to carefully weigh all of the evidence, taking into account prevailing rules for the weight that should be applied for various classes of evidence to find the most probable location where the original surveyor actually placed a corner marker. In this process, it should always be remembered that the goal of a retracement survey is, as the name implies, retracement. That process is frequently described as "following in the footsteps of the original surveyor."

When a retracement survey fails to discover acceptable evidence of the exact, original location of a property corner and discrepancies exist in the record dimensions to other monuments, the process of apportionment should be applied to find the most probable location for the lost corner. Apportionment is a method of distributing excess or deficiencies between two corners so that a proportioned amount, based on record distances, is given to the distances from the found corner to the lost corner. It is emphasized that apportionment is a last resort measure. It should be utilized only when there is absolutely no evidence of the original location of a boundary corner.

Apportionment is appropriate only when retracement fails to yield the position of a corner (Griffin 1960).

For surveys in the Public Land Survey System, the apportionment process that should be followed is described in the following section. For other types of subdivisions, the process is similar. For example, if monuments are found for block corners in a subdivision, while the individual lot corners needed for a survey are lost, any excess or deficiency between the record and measured distances between the block corners should be distributed proportionally, based on the record distances, throughout the line within each block (Figure 6.1). It should be noted that apportionment should be applied only when dealing with parcels that are created simultaneously, such as in the case with subdivisions, as opposed to situations where there may be seniority of title based on date of creation. Several conditions should be observed when applying apportionment to a subdivision. One is that no portion of excess or deficiency should be allocated to the streets within a tract. The apportionment should be confined to a block. A second condition applies to subdivisions with plats having all lots dimensioned except for an end one that is undimensioned. Based on an early common practice associated with imprecise measurement where any excess or deficiency is placed in the remnant lot, courts in some jurisdictions have held that this is evidence that the subdivider intended any discrepancy to be borne by the irregular end lot.

When evaluating various categories of evidence found in a retracement survey, conflicts between the calls in the description often develop. Therefore, certain rules have evolved for weighing the various types of evidence. For example, distances and directions between found original measurements take priority over plat distances and directions. Generally, in the absence of a clear intention to the contrary, the order of priority of calls is as follows:

1. Calls for natural monuments such as rivers, creeks, and mountains
2. Calls for manmade monuments such as concrete monuments
3. Calls for boundaries of adjoining lands
4. Calls for course
5. Calls for distance
6. Calls for area

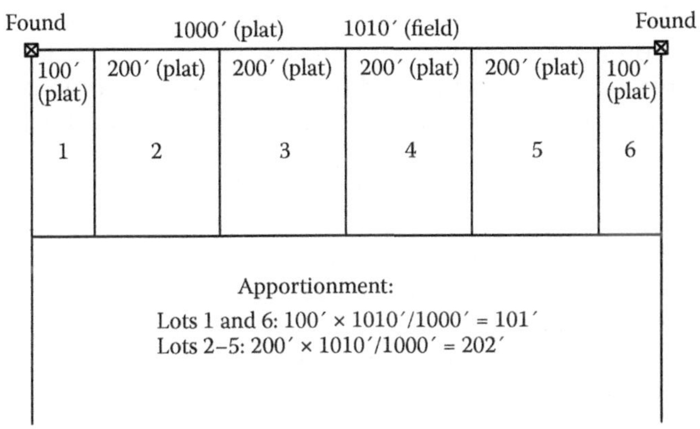

FIGURE 6.1
Apportionment in subdivisions. The excess found between the original block corner monuments should be distributed proportionally to the front lot boundaries since no lot corner monuments were found.

6.2.3 Retracement of Public Land Surveys

With retracement surveys of Public Land Survey sections, procedures prescribed in the *Manual of Surveying Instruction* (Bureau of Land Management 2009) should be followed. As part of this process, the original field notes and the plat should be examined and a thorough search should be made for evidence described in those documents as well as for evidence associated with subsequent surveys. When surveying aliquot (fractional) parts of public land sections, the center of the section should be located at the intersection of lines connecting opposing quarter corners.

If acceptable evidence of an original corner monument or accessories to that corner, such as witness trees, or other original evidence such as topographic calls is found, the corner is considered an *existent corner.* If no original evidence for the corner is found but acceptable evidence of the corner location such as subsequent surveys or personal testimony (parol evidence) is found, the corner is considered an *obliterated corner.* If neither acceptable original evidence nor subsequent evidence regarding the original location of the corner is found, the corner is considered a *lost corner,* and proportionate measurement should be used to find the most probable location for the corner. That process should be applied only when all means of determining the original position of the corner have been exhausted.

There are two methods prescribed for proportionate measurement for public land survey corners. These are single proportionate measurement and double proportionate measurement. The choice between them will depend on the type of corner. Both processes involve survey location of accepted corners in all controlling directions, two directions for single proportionate measurement and four directions for double proportionate measurements. In both cases, each known corner is given a controlling weight inversely proportional to its distance from the lost corner. The corners should be the closest known corner to the lost corner in each direction, but the control corners may not be beyond the township boundaries. If the township boundary is reached before finding a known corner in any direction, a corner on the township boundary must be reestablished before proceeding.

Single proportionate measurement should be applied to measurements made between two known corners where those corners alone control the direction of the line. Examples are quarter corners and intermediate corners on township lines. This process requires the location by survey of two known corners, one on either side of the lost corner. With a configuration such as illustrated in Figure 6.2, coordinates for the lost corner may be calculated as follows:

$$Y_3 = Y_1 + (Y_2 - Y_1)\frac{\text{record dist}_{1-3}}{\text{record dist}_{1-2}} \tag{6.1}$$

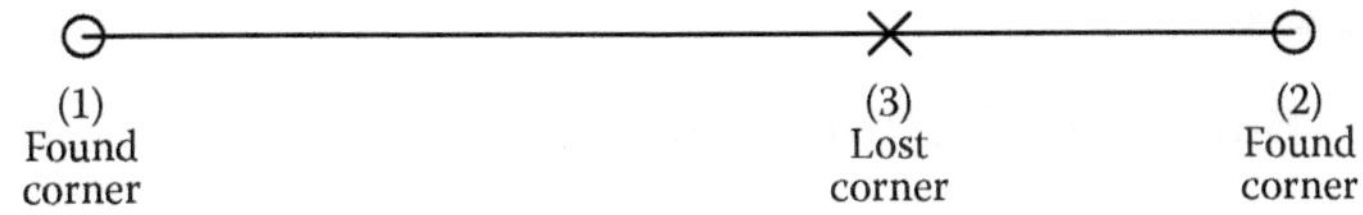

FIGURE 6.2
Single proportionate measurement.

$$X_3 = X_1 + (X_2 - X_1)\frac{\text{record dist}_{1-3}}{\text{record dist}_{1-2}} \tag{6.2}$$

Double proportionate measurement is applicable to measurements made between four known corners, two each on intersecting meridian and latitudinal lines. Examples are section corners in the interior of a township and township corners. This process requires the location by survey of four known corners, one on each side of the lost corner. With a configuration such as illustrated in Figure 6.3, coordinates for the lost corner may be calculated as follows:

$$Y_5 = Y_1 + (Y_3 - Y_1)\frac{\text{record dist}_{1-5}}{\text{record dist}_{1-3}} \tag{6.3}$$

$$X_5 = X_2 + (X_4 - X_2)\frac{\text{record dist}_{2-5}}{\text{record dist}_{2-4}} \tag{6.4}$$

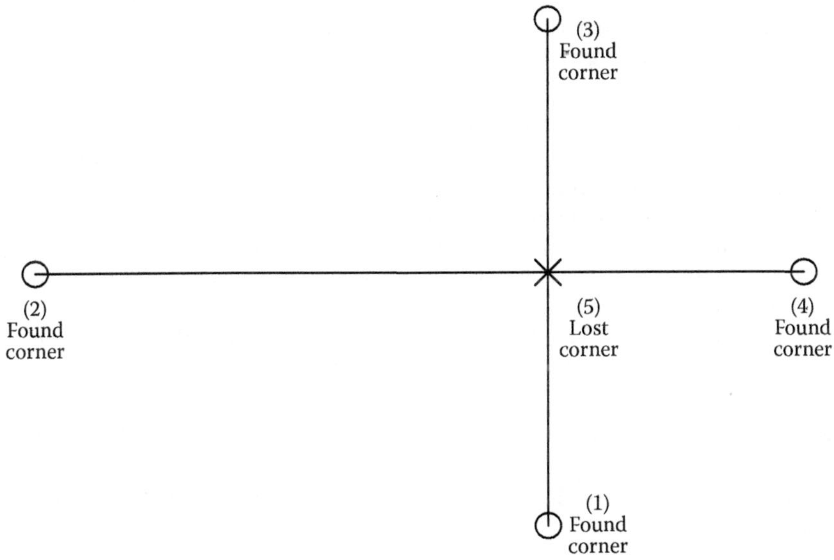

FIGURE 6.3
Double proportionate measurement.

6.3 Water Boundary Delineation

6.3.1 Origin and Evolution of Water Boundary Definitions

Water boundaries are ideal natural boundaries since they are easily recognized and relatively easy to defend. Thus, they have been among the most widely used boundaries throughout history. Yet, despite their long-established and widespread use, water boundaries are probably among the most frequently and bitterly contested of boundaries in today's society. This is primarily due to the fact that the land–water interface is dynamic and constantly changing due to tides and/or meteorological conditions. In addition, the shoreline in many areas constantly changes due to erosion and accretion caused by waves and currents. Therefore, unlike most other boundaries that are two-dimensional and stationary, a third dimension—height—and a fourth dimension—time—must be considered when dealing with water boundaries. Consequently, unique laws and techniques have developed over the years for defining and locating water boundaries (Cole 1997).

One class of water boundaries that have been especially contentious are those representing the boundary of publicly owned waters. Such public ownership is based on a long-established policy of considering the seas and other navigable waters as a public commons, often called the Public Trust Doctrine. That doctrine has roots as far back as the early Roman Empire. This may be illustrated by the Roman Civil Code, the *Institutes of Justinian*, prepared under the direction of Emperor Justinian in 529 AD, which stated as follows (Sandars 1874):

> Section 1, Book II, Title I: By the law of nature these things are common to mankind—the air, running water, the sea, and consequently the shores of the sea. No one, therefore, is forbidden to approach the seashore, provided that he respects habitations, monuments, and buildings, which are not, like the sea, subject only to the law of nations… The seashore, that is the shore as far as the waves go at furthest, is considered to belong to all men.
>
> Section 2, Book II, Title I: All rivers and ports are public; hence the right of fishing in a port or in rivers, is common to all men.

Early Spanish civil law in *Las Siete Partidas* (Scott 1931), written and compiled in the 13th century under the order of Alfonso the Wise of Castille, more or less tracked the Roman *Institutes of Justinian*, as illustrated by the following cite:

> The things which belong in common to the creatures of this world are the following, namely; their, the rain-water, and the sea and its shores, for every living creature can use each of these things, according as it has

> need of them. For this reason every man can use the sea and its shore for fishing or for navigation, and for doing everything there which he thinks may be to his advantage.*

English common law also considered tidal waters to be sovereign, as indicated by the following quote from the early case of *Attorney General v. Richards*†

> ...the sea and sea coasts, and as far as the sea flows and reflows, between the high and low-water marks, and all the ports and havens of the kingdom, belong to his Majesty...

Regarding nontidal waters, English common law did not consider them as sovereign, as demonstrated by the following cite from Lord Chief Justice Mathew Hale in his *De Jure Maris* (Hale 1666):

> Fresh rivers of what kind soever, do of common right belong to the owners of the soil adjacent; so that the owners of the one side have, of common right, the property of the soil and consequently the right of fishing, usque filum aquae [to the middle of stream]; and the owners of the other side the right of soil or ownership and fishing unto the filum aquae on their side. And if a man be owner of the land on both sides, in common presumption he is the owner of the whole river, and hath the right of fishing according to the extent of his land in length....

Nevertheless, navigable nontidal rivers were considered to be public highways, even though privately owned, as illustrated in the following clause from the same writing:

> There be some streams or rivers, that are private not only in propriety or ownership, but also in use, as in little streams and rivers that are not a common passage for the king's people. Again, there be other rivers, as well fresh as salt, whether they flow and reflow or not, are prima facia publici juris, common highways for man or goods or both from one inland town to another. Thus the rivers of Wey, of Severn, of Thames, and divcers others, as well above the bridges and ports as below, as well above the flowings of the sea as below, as well where they are become to be of private propriety as in what parts they are of king's propriety, are publick rivers juris publici.

In the United States, the Public Trust Doctrine is manifested in laws holding that the marginal territorial seas along the coastline are public trust lands owned by the public of each state, based on the following U.S. Supreme Court opinion.

---

* Partida 3, Title 28, Law 4

† Anstruther's Rep. 603

> When the revolution took place the people of each state became themselves sovereign and in that character hold the absolute right to all their navigable waters in the soils under them for their own common use...*

The question of whether or not all tidally affected waters, regardless of their navigability-in-fact, are considered as sovereign is one that has been the subject of considerable dispute. In some states, public ownership is considered to extend to all submerged lands subject to the ebb and flow of the tide, regardless of actual navigability-in-fact. In Louisiana, Maryland, Mississippi, New Jersey, New York, and Texas, all waters subject to tidal ebb and flow are considered to be sovereign, while in California, Connecticut, Florida, North Carolina, and Washington, whether tidally affected waters are sovereign has traditionally been based on navigability-in-fact of the waters (Maloney and Ausness 1974).

As an illustration of this distinction, the preponderance of historic Florida case law appears to support the navigability-in-fact approach. For example, *Clement v. Watson*† states that "waters are not under our law regarded as navigable merely because they are affected by the tide." Other cases reflect this same position. Furthermore, the navigability-in-fact position is clearly implied in Section 177.28 of the Florida Statutes, which states that the "...mean high water line along the shores of land immediately *bordering on navigable waters* is recognized and declared to be the boundary between the foreshore owned by the state in its sovereign capacity and upland subject to private ownership" (emphasis added).

As opposed to states such as Florida with navigability-in-fact case law, the State of Mississippi has been generally considered to be an "ebb and flow" state. In a recent challenge (*Cinque Bambini Partnership v. State of Mississippi*‡) in that state as to whether all tidally affected waters are sovereign regardless of navigability-in-fact, both the chancery court and the Mississippi Supreme Court decided in favor of ebb and flow. Salient excerpts from the colorful State Supreme court opinion are as follows:

> The early federal cases refer to the trust as including all lands within the ebb and flow of the tide.
>
> ...it is our view that as a matter of federal law, the United States granted to this State in 1817 all lands subject to the ebb and flow of the tide and up to the mean high water level, without regard to navigability.
>
> Yet so long as by unbroken water course—when the level of the waters is at mean high water mark—one may hoist a sail upon a toothpick and without interruption navigate from the navigable channel/area to land, always afloat, the waters traversed and the lands beneath them are within the inland boundaries we consider the United States set for the properties granted the state in trust.

* *Martin v. Waddell*, 41 U.S. (16 Pet.) 367 (1852)
† Fla, 58 So. 25 (1912)
‡ Miss, 491 So. 2d 508 (1986)

The case was appealed to the U.S. Supreme Court (*Phillips Petroleum Co. v. Mississippi**), which concurred with the state court in ruling that all coastal states received all lands over which tidal waters flow and that the public of Mississippi still owns such land. The majority opinion by Justice O'Connor noted that this ruling "will not upset titles in all coastal states [since it] does nothing to change ownership rights in states which previously relinquished a public trust claim to tidelands such as those at issue here." Whether or not the ruling has an effect in states, such as Florida, which have established navigability-in-fact case law, is yet to be tested in a judicial proceeding (Cole 1991).

Regarding nontidal waters, in 1876, in the case of *Barney v. Keokuk*,† the U.S. Supreme Court ruled that state title in navigable waters extended to inland, nontidal waters as well as tidal waters, with the following words:

> The confusion of navigable with tide water, found in the monuments of the common law, long prevailed in this country, notwithstanding the broad differences existing between the extent and to of the British island and that of the American continent…And since this court…has declared that the Great Lakes and other navigable waters of the country, are, in the strictest sense, entitled to the denomination of navigable waters, and amenable to the admiralty jurisdiction, there seems to be no sound reason for adhering to the old rule as to the proprietorship of the beds and shores of such waters. It properly belongs to the States by their inherent sovereignty…

Thus, many of the states, in their sovereign capacity, claim title to the beds under navigable waters regardless of whether or not the waters are tidally affected, although some states do not claim public ownership of the submerged lands under nontidal water. Nevertheless, in the states that do not claim sovereign ownership of such waters, they are considered to be public highways and subject to a public easement.

In nontidal waters, navigability for title purposes generally is a question of navigability-in-fact. In Florida, for example, case law (*Odom v. Deltona*‡) offers specific clarification to that state's definition in nontidal waters. In that case, the court held that "Florida's test for navigability is similar, if not identical, to the Federal Title Test." The Federal Title Test was defined as being "…based on the body's potential for commercial use in its ordinary and natural condition."

Further understanding of the Federal Title Test may be obtained from the following early statement of this test (The Daniel Ball§):

> …and they are navigable in fact when they are used or susceptible of being used in their ordinary condition, as highways for commerce, over which trade and travel are or may be conducted in the customary modes of trade and travel over water.

* 484 US 469 (1988)
† 94 U.S. 324 (1876)
‡ Fla, 341 So 2d 977 (1976)
§ 77 US 557 (1870)

Thus, the definition of water boundaries varies with whether or not it is tidal and whether or not it is navigable. In addition, definitions for such boundaries often vary with the jurisdiction in which the property is located. Therefore, all of these factors must be considered when surveying such boundaries.

6.3.2 Boundaries of Tidal Public Trust Waters

When dealing with tidally affected waters, a threshold question is the definition of navigability. Under the Submerged Lands Act (43 U.S.C., s1301-1 1970), the coastal waters adjacent to each of the coastal states are owned by the public of those states. In addition, all natural interior tidal waters within their boundaries, unless validly conveyed, are considered to have been conveyed to the public of each of the U.S. coastal states with statehood. Furthermore, based on the previously mentioned *Phillips Petroleum v. Mississippi** case, this is true regardless of their navigability-in-fact.

Dealing with the boundaries between the submerged lands of such tidally affected public trust waters and the bordering uplands, subject to private ownership, three approaches to tidal boundaries prevail in the United States. Those approaches are described in the following three sections and include the Anglo-American Common Law, the North-Atlantic Low Water States, and the Civil Law.

6.3.2.1 Anglo-American Common Law

The most prevalent of approaches dealing with the private/public boundaries of tidally affected public trust waters in the United States is that taken by Anglo-American Common Law. For states following this law, that boundary is considered to be the mean high water line, which represents the average upper reach of the daily tides. The common definition of that line is that formed by the intersection of the rising coastline and the average of the tidal high waters over a period defined by the astronomical forces affecting the tides. Thus, it uses the average of observed tidal high waters over a complete tidal epoch of 19 years† or the equivalent based on a correlation between short-term observations at a local tide station and a control station with established 19-year mean values.

Average tidal data has a long history of use. For example, probably the earliest mention of using tides to determine the coastal boundary was in the late 1400s by Thomas Diggs, a surveyor, engineer, and lawyer, in a book entitled *Proofs of the Queen's Interest in Land left by the Sea and the Salt Shores Thereof.* That approach was further amplified a century or so later by Lord Chief Justice Matthew Hale, in a book entitled *De Jure Maris*, which declared

* 484 US 469

† 19 years represents the period during which all of the recognized harmonic cycles affecting the tides occur, including the 18.6-year cycle associated with the regression of the moon's nodes.

that the foreshore, the area that is overflowed by "ordinary tides or neap tides, which happen between the full and change of the moon," belongs to the crown. A latter British case, *Attorney General v. Chambers*,* ruled that the boundary was to be found by "the average of medium tides in each quarter of a lunar evolution during the year."

In the United States, the landmark case of *Borax Consolidated Ltd. v. City of Los Angeles*† set the standard for the United States as follows:

> In view of the definition of the mean high tide...and the further observation that...there should be "a periodic variation in the rise of water above sea level having a period of 18.6 years"...in order to ascertain the mean high tide line with requisite certainty..."an average of 18.6 years should be determined as near as possible."

The basis for the 18.6-year period of averaging mentioned in the previous cite is that it includes at least one occurrence of all of the astronomical cycles that serve as constituents of the observable daily tidal cycles. These include forces such as those associated with the daily orbits of the moon about the earth, the rotation of the earth in relation to the sun, the monthly lunar and annual solar declination cycles, and the 18.6-year cycle for the regression of the moon's nodes.‡ In practice, a 19-year averaging period is used for inclusion of an integral number of the annual cycle associated with the declination of the sun.

As may be seen from the definition as it evolved, the mean high water line represents an attempt to define the upper reach of the daily tide as the boundary between publicly owned submerged lands and uplands subject to private ownership. Since the upper reach of the tide varies from day to day, this definition is an attempt at a compromise line reflecting the *average* daily upper reach of the tides. The result is in a line that is exceeded by the high tide on approximately one-half of the tidal cycles (Figure 6.4).

Case law in the various coastal states has generally followed the English common law and its updated definition as put forth in the Borax decision. Sixteen states (Alabama, Alaska, California, Connecticut, Florida, Georgia, Maryland, Mississippi, New Jersey, New York, North Carolina, Oregon, Rhode Island, South Carolina, Texas, and Washington) have followed this course (Maloney and Ausness 1974; Cole 1997). Some states have codified their common law on this subject. As an example, in Florida, the Coastal Mapping Act of 1974 (Chapter 177, Part II, Florida Statutes) declares that "mean high water line along the shores of lands immediately bordering on navigable waters is recognized and declared to be the boundary between the foreshore owned by the State in its sovereign capacity and upland subject to

* *A*17 Eng. Rul. Cas. 555 (1889)

† *Borax Consolidated Ltd. v. City of Los Angeles,* 296 U.S. 10 (1935)

‡ Regression of the moon's nodes refers to the movement of the intersection of the moon's orbital plane and the plane of the Earth's equator, which completes a 360° circuit in 18.6 years.

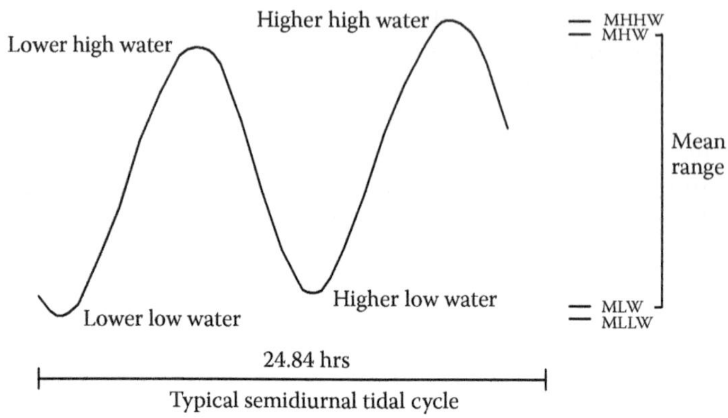

FIGURE 6.4
Average tidal cycles.

private ownership." That statute further defines the mean high water line using the Borax concept as follows:

> Mean high water means the average height of the high waters over a 19-year period. For shorter periods of observation, mean high water means the average height of the high waters after corrections are applied to eliminate known variations and to reduce the result to the equivalent of a mean 19-year value. (Section 177.27 (14), Florida Statutes)

In addition to the predictable periodic variation in tidal heights occurring within a 19-year period, there are also nonperiodic changes due to "glacial-eustacy, thermal volumetric changes, vertical land movements, and both climatological and oceanographic trends" (Hicks et al. 1983). Therefore, in the United States, tidal datum planes, such as mean high water, are calculated using a specific 19-year epoch. Periodically, a new national epoch is adopted by the NOAA after significant change has occurred. At the time of this writing (2015), the National Tidal Datum Epoch is 1983–2001.

The location of a boundary under this approach requires the determination of the local elevation of mean high water since there may be considerable geographic variation in the elevation of that datum. This may be accomplished by use of an existing tide station such as those created by NOAA along the nation's coastline at which tidal data based on the current tidal epoch have been established. If such a station does not exist in the immediate vicinity of the need, local tides should be observed simultaneously with observations at such an established tide station.

Depending on the distance and hydrographic conditions between the site and the control station, the recommended duration of such observations may vary from a few days for a nearby control station to a year or more for situations where the control station is a considerable hydrographic distance away. The observations may then be correlated using a simultaneous comparison process (Cole 1997). Once the local elevation is established, that contour may be traced along the shoreline to delineate the mean high water line.

6.3.2.2 North-Atlantic Low Water States

Five Northeastern States (Maine, Massachusetts, Delaware, Pennsylvania, and Virginia) have adopted *mean low water* as the coastal boundary. As with mean high water, mean low water is defined as the average of all of the tidal low waters over a 19-year tidal epoch. That line allows greater access and usage of the shoreline by riparian or littoral owners. In the New England states, this policy is based on a Massachusetts colonial ordinance of 1641–1647 as follows:

> …in all creeks, coves and other places, about and upon salt water where the Sea ebbs and flows, the Proprietor of the land adjoining shall have proprietie to the low water mark where the Sea doth not ebb above a hundred rods, and not more wheresoever it ebbs farther.

The language of the ordinance has generally been interpreted as defining the waterward boundary of lands bordering on tidal waters as the mean low water line unless that line is 100 rods distant from the mean high water line. If there is more than a 100-rod distance between the two lines, then the boundary is a line 100 rods seaward of the mean high water line. Therefore, ownership of riparian or littoral lands under this legal system includes the tidal flats often found in such areas.

States adopting this approach for their coastal boundaries typically reserve a public easement up to the mean high water line even though public ownership extends only to the mean low water line. Such a reservation is illustrated by the following (*Tinicum Fishing Co. v. Carter**):

> The title of a riparian owner extends to low-water mark, not absolutely in tidal streams, but subject to the public right of passage when the tide is high.

The process for calculation of the tidal datum used for those states recognizing the mean low water line as the boundary is essentially the same as that recognizing the line of mean high water. The only difference is that the average of the low waters of the daily tidal cycles over a 19-year tidal epoch is used rather than the high waters.

6.3.2.3 Civil Law

The civil law approach to defining a coastal boundary prevails in areas where the land title has its roots in a grant from a sovereign power where civil law, such as the previously mentioned Roman *Institutes of Justinian*, prevailed. Possibly due to the civil law code being developed in an area in the

* 61 PA 21 (1869)

Mediterranean with little daily tidal range as opposed to the greater daily tidal range in the British Isles, Roman civil law defines the coastal boundary in terms of seasonal water levels rather than in terms of daily tide levels. As an example, a translation of a portion of the Roman *Institutes of Justinian* (Sandars 1874) follows:

> The sea-shore, that is, the shore as far as the waves go at furthest, was considered to belong to all men.... The sea shore extends as far as the greatest winter floods run up.

The previously mentioned U.S. case of *Borax Consolidated v. City of Los Angeles* clearly distinguishes between the Anglo-American common law and the civil law as defined in the Roman Civil Code: By the civil law, the shore extends as far as the highest waves reach in winter. But by the common law, the shore "is confined to the flux and reflux of the sea at ordinary tides."

As an example of the use of the civil law definition, in Louisiana, the waters of the Gulf of Mexico, including the shores of the Gulf, are in public ownership and are considered to extend to the "highest tide during the winter season" (Article 451. LA Civil Code and LA Rev. Stat. 49:3). Interestingly, this definition results in a lower boundary than the mean high water line since the high tide during the winter in the northern Gulf of Mexico is usually less than the high tide during other parts of the year because of the prevailing northern winds in that area during winter. The State of Hawaii also appears to follow the civil law approach, with a coastal boundary defined as "the upper reach of the wash of the waves" (Maloney and Ausness 1974).

Another example is the State of Texas, which has recognized the civil law definition in areas of the state with origins of land title in Spanish or Mexican land grants. The Commonwealth of Puerto Rico also follows the civil law. For both Texas and Puerto Rico, it has been held that the limit of ownership is controlled by old Spanish civil law contained in *Las Siete Partidas*, written in the 13th century, where the boundary is defined as follows:

> ...e todo aquel lugar es llamado ribera de la mar quanto se cubre el agua della, quanto mas crece en todo el año, quier en tiempo del invierno o verano... (Partida 3, Title 28, Law 4)

The correct interpretation of this clause has been the subject of considerable judicial debate. Possibly, the leading English translation of these laws (Scott 1931), translates it as follows:

> ...and all that ground is designated the shore of the sea which is covered with the water of the later at high tide during the whole year, whether in winter or in summer.

The intent of *Las Siete Partidas* regarding the location of the upper limit of the public commons may be better understood by seeing the entire Scott translations of the two laws (with emphasis added) pertinent to the previous clause, as follows:

> *The things which belong in common to the creatures of this world are* the following, namely; the air, the rain-water, and *the sea and its shores,* for every living creature can use these things, according as it has need of them. (Partida 3, Title 28, Law 3)
>
> Every man can build a house or a hut on the sea shore which he can use whenever he wishes; and any person can erect there another edifice for his own benefit, provided the common custom of the people is not violated; and he can construct galleys and any other kind of ships and dry nets there and make new ones if he desires to do so; and so long as he is working there or is present no one else should molest him so that he cannot use and be benefited by all these things, and by others like them, in the manner aforesaid; and *all that ground is designated the shore of the sea which is covered with the water of the latter at high tide during the whole year, whether in winter or in summer.* (Partida 3, Title 28, Law 4)

It may be readily seen that the Spanish civil law differs from the Roman in that rather than use of the phrase "as far as the highest waves reach in winter" as in the Roman *Institutes of Justinian,* this Scott interpretation of *Las Siete Partidas* calls for the line where the land is covered by the high tide whether in winter or summer (*invierno o verano*). Use of the words *high tide* suggests a boundary determined by use of tidal data as opposed to the use of water levels caused by metrological events. That interpretation is credible since, unlike the Mediterranean coast of Italy (home of the Roman law), with an almost imperceptible tidal range, Spain has coastlines on both the Mediterranean Sea and Atlantic Ocean. Therefore, the authors of *Las Siete Partidas* would have been far more aware of the prominent daily tidal cycle and its effect on the shore than would the authors of the Roman *Institutes of Justinian.* Yet, considering that this section of the code also states that the public commons include the shore where fishermen have the right to dry their nets and construct huts, this suggests a boundary considerably higher than the mean high water line of the Anglo-American common law.

One judicial interpretation of *Las Siete Partidas* was made by the Supreme Court of Texas in *Luttes v. State** in 1950. In that trial, testimony provided various interpretive translations that suggest that the proper meaning is alternatively the highest swell of the year, the highest tide of the year, or an average high tide. Much of the deliberation centered about the historic meaning of the Spanish verb *crecer,* or "to grow, expand, or rise up," and whether it means "rise of the water of the sea by effect of the tide (*marea*)." The Court concluded that the language of *Las Siete Partidas* implied an average tide and that the applicable rule is that of the "average of highest daily water computed over or corrected to the regular tidal cycle of 18.6 years. This means

* 324 S.W. 2d 167

in substance mean high tide." That language was clarified in a response to a motion for rehearing with the statement that it was the court's intention that the line was actually that of mean higher high tide,* as distinguished from the mean high tide of the Anglo-American law. That clarification was apparently in recognition that the prevailing tide along the South Texas coast is diurnal, with only one tidal cycle per day most of the month so that the average of the daily high tides is essentially the mean higher high water, which is the average of the highest of tides that occur each day.

A considerably different interpretation prevails in Puerto Rico, which also follows the civil law as prescribed in *Las Siete Partidas*. The decision in case of *Rubert Armstrong v. E.L.A*† held that Puerto Rico has

> el mismo criterio de la ley de Costas de España, de que el limite de la zona maritime-terrestre será el que abarque un mayor ámbito físico, hasta donde alcancen las olas en los mayores temporales o, cuando lo supere, la línea de pleamar máxima viva equinoccial.

This may be loosely translated as having "the same criteria as the coastal law of Spain, with the coastal boundary including the greatest physical environment, where the waves of the greatest storms reach, or when it is higher, the line of the equinoctial spring tide."‡

Furthermore, territorial guidelines for determining coastal boundaries in Puerto Rico§ state:

> Para determinar el alcance tierra adentro del criterio mareal se pueden utilizar los indicadores bióticos o el alcance de la pleamar máxima viva equinoccial....

which may be loosely translated as "for determining the landward reach of the tides, you can use the biological indicators *or* the reach of the *equinoctial spring tide*" (emphasis added).¶

---

* Mean higher high water is the average of the higher of the two daily tidal high waters each day over a 19-year tidal epoch.

† 97 D.P.R. 588 (1969), el Tribunal Supremo de Puerto Rico

‡ Spring tides are those experienced twice a month near the time of the new and full moon when the earth, moon, and sun are in a line. Tidal ranges for spring tides are considerably greater than average tides in most areas due to the gravitational forces of the moon and sun pulling in unison. Equinoctial spring tides result twice annually when spring tides occur near the times of the vernal or autumnal solar equinox. At those times, the sun is over the equator and the paths of the sun and moon are in closest alignment resulting in tidal ranges greater than average spring tides.

§ Section 3.2.1.a (2), Manual de Procedimientos para el Deslinde del Limite Interior Tierra Adentro de los Bienes de Dominio Publico Maritimo Terrestre, DRNA (1999)

¶ In practice, in Puerto Rico, the boundary is usually determined by vegetative and geological features reflecting highest water levels, including storm surges. Tests of such a line show that it is higher than either the equinoctial spring tide or the highest astronomic tide and therefore reflects a line created by higher water levels such as those associated with storm events (Cole 2011).

Another approach that has been suggested in Puerto Rico (Cole 2011) is the use of the *highest astronomical tide,* which is the elevation of the highest predicted astronomical tide expected to occur at a specific location over a tidal epoch of 19 years. This occurs when the major astronomic cycles that cause the observable daily tide are all "in phase" and at their maximum. This latter datum would more or less represent the most landward tide possible under average weather conditions and can be determined with knowledge of the harmonic constituents of the tide at a specific location.

All of the alternatives discussed previously for compliance with the original intention of the Spanish code of *Las Siete Partidas* offer the objectivity and mathematical certainty of tidal boundaries. Yet, it would seem that the interpretation offered by the Luttes case in Texas does not meet the definition of including both the sea and the shore as part of the public commons, while either the equinoctial spring tide or the highest astronomical tide does include at least a portion of the shore and thus best meets the criteria stated in the Spanish code.

6.3.3 Boundaries of Nontidal Public Trust Water

With the lack of the predictable rise and fall of water level observed in tidal waters, different boundary definitions apply in nontidal waters than in tidal waters. Rather than using the level of the water itself to determine the boundary as with tidally affected waters, the laws in the various states have generally called for the use of surrogate evidence and defined the boundary of publicly owned nontidal waters as that determined by various physical indicators of the ordinary water levels. The laws of some states define the boundary of such waters as the ordinary high water mark, while others, as the ordinary low water mark.* Florida is a typical example of states holding the ordinary high water mark as the boundary, with case law such as the following (*Tilden v. Smith*†):

> High-water mark, as a line between a riparian owner and the public, is to be determined by examining the bed and banks, and ascertaining where the presence and action of the water as so common and usual, and so long continued in all ordinary years as to mark upon the soil of the bed a character distinct from that of the banks, in respect to vegetation as well as respects the nature of the soil itself. High-water mark means what its language imports—a water mark.

* States that have adopted the ordinary high water mark include Alaska, Florida, Idaho, North Carolina, New Hampshire, Oregon, Washington, and Wisconsin (Coastal States Organization 1997). States that have adopted the ordinary low water mark include Alabama, Delaware, Indiana, Louisiana, Massachusetts, Minnesota, Missouri, Montana, Pennsylvania, South Dakota, Tennessee, and West Virginia (78 AM JUR. 2d Water 386, 1975).

† Fla, 113 So 708 (1927)

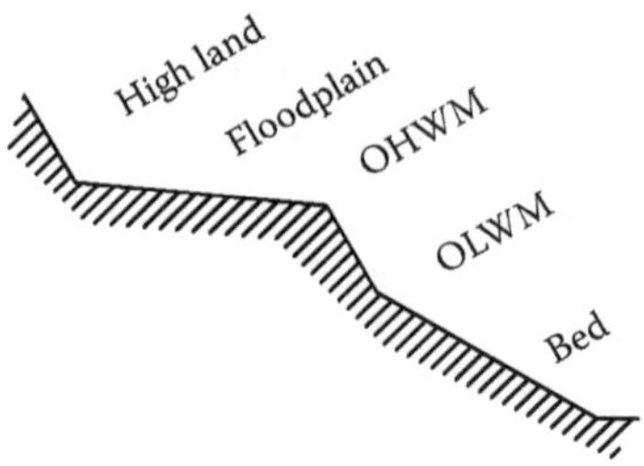

FIGURE 6.5
Generalized shoreline profile.

Examples of the use of the ordinary low water mark may be found in case law in the State of Pennsylvania such as the following (*Freeland v. Pa. Railroad Co.**):

> (Land) bounded by a large navigable river, vests in the owner the right of soil to ordinary low water mark of the stream, subject to the public right of passage for navigation, fishing, etc., in the stream, between ordinary high and ordinary low watermark.

Evidence for nontidal water boundaries—Considering the previous citation, evidence that should be evaluated for determining the appropriate location of a nontidal water boundary may be classified in four categories: geomorphology, botany, soil, and hydrology (Cole 1997).

Regarding the location of this boundary, the *Manual of Surveying Instruction* (Bureau of Land Management 2009) provides the following guidance:

> The most reliable indicator of mean (ordinary) high water elevation is the evidence made by the water's action at its various stages, which are generally well marked in the soil. In timbered localities, a very certain indication of the locus of the various important water levels is found in the belting of the native forest species.

Geomorphological features—These include features indicative of the natural limits of water bodies such as escarpments and natural levees. The margin of most water bodies tends to form a similar profile (Figure 6.5), although the width and slope of the floodplain, foreshore between the ordinary high and low water elevations, and bed may vary considerably with different water bodies. In riverine systems, there may be considerable variation due to the meandering of the river, including the existence of natural flood-formed levees at the waterward edge of the floodplain.

Based on the generalized profile (Figure 6.5), the objective of an ordinary high water determination is to locate the dividing line between the floodplain and the bed of the water body. The *Manual of Surveying Instruction*

* 197 Pa. 529, 47 A. 745 (1901)

(Bureau of Land Management 2009) provides guidance regarding the use of geomorphological features for this determination as follows:

> Mean (ordinary) high water elevation is found at the margin of the area occupied by the water for the greater portion of each average year. At this level a definite escarpment in the soil is generally traceable, at the top of which is the true position for the meander line. A pronounced escarpment, the result of the action of storm and flood waters is often found above the principal water level, and separated from the latter by the storm or flood beach. Another, less evident, escarpment is often found at the average low water level, especially of lakes, the lower escarpment being separated from the principal escarpment by the normal beach or shore. While these principles properly belong in the realm of geology, they should not be overlooked in the survey of a meander line.

As mentioned in the *Manual of Surveying Instruction*, escarpments are also sometimes found at the upper extreme of the floodplain. In rivers, this may be some distance from the ordinary high water mark. The more significant escarpment for boundary determination would be found near the open water.

Another geomorphological feature often found near the ordinary high water line in rivers is a natural flood-formed levee. *Natural levees* are ridges that parallel a river course. They are the highest near the river and slope gradually away from it. In larger rivers, they may be several feet high and a mile or more in width. In other rivers, they may be only a few inches high and a few feet wide. Levees owe their greater height near the stream channel to the deposition of material when the river overspreads its banks during flood events. Therefore, the ordinary high water level is usually on the river side and below the crest of such features. Erosional features, such as escarpments, may often be found on the waterward side of levees (Cole 1997).

Vegetation—This category includes evidence such as the lower limit of terrestrial vegetation and areas where vegetation has been wrested away by wave action or stream currents. The *Manual of Surveying Instruction* gives the following directions regarding the use of such evidence:

> Where native forest trees are found in abundance bordering bodies of water, those trees showing evidence of having grown under favorable site conditions will be found belted along contour lines. Certain mixed varieties common to a particular region are found only on the lands seldom, if ever, overflowed. Another group is found on the lands which are inundated only a small portion of the growing season each year, and indicate the area which should be included in the classification of the uplands. Other varieties of the native forest trees are found only within the zone of swamp and overflowed lands. All timber growth normally ceases at the margin of permanent water.

The rationale for the use of such evidence is fairly evident. It has been observed and well established that many forms of plant life are distinctly related to the amount and duration of water to which they are subjected. Therefore, it is reasonable that with knowledge of the water tolerances of the plant life for a particular geographical area, patterns of such growth may offer significant assistance in the location of the ordinary high water line.

Change in the character of the soil—This includes evidence such as differences in organic content due to leaching and the landward limit of stratified beach deposits where waterward transport of eroded shoreline has resulted in a natural sorting of material. That transportation generally results in a systematic decrease in average grain size when proceeding waterward. This type of evidence has limited application in rivers due to seasonal flooding and the meandering process but may be helpful in lakes (Cole 1997).

Such definitions sound fairly straightforward but often result in problems when being applied. There is often conflicting physical evidence indicating different stands of water. Further, evidence may have been obliterated by shoreline development or higher than ordinary water levels. Evidence of the ordinary low water mark is especially difficult to find since that elevation may be underwater much of the time and evidence obliterated by higher stands of water. To further complicate this challenge, there is little judicial guidance from case law as to how to find the low water mark. As a result of these problems together with the intense land use and the high monetary and environmental value associated with waterfront property today, disputes may arise. Several very bitter boundary disputes have occurred recently, where the horizontal differences in the two asserted positions of the ordinary high water mark have been several miles apart, with claims of supporting physical evidence found at both positions.

Hydrographic records—Traditionally, in nontidal waters, the courts have allowed the use of botanical and geological evidence, as evidenced by the previous decisions, but not allowed the use of mathematical averaging of water levels. This is typified by the court's decision in *Kelly's Creek and N.W.R. Co. v. United States*:

> The high water mark is not to be determined by arithmetical calculation; it is a physical fact to be determined by inspection of the river bank.

Recently, however, there has been an apparent trend to place more reliability on water level records, possibly due to the growing need for the precision, repeatability, and lack of ambiguity that results from a mathematical solution (Cole 1997). Typical of this are two Florida cases, *U.S. v. Parker* and *U.S. v. Joder Cameron*. The court in the Cameron case found as follows:

> There is no logical reason why a fourth approach to determining the line or ordinary high water may not consist of comparing reliable water stage and elevation data. Indeed, for a body of water whose levels fluctuate considerably with changes in climate, accurate water stage and elevation data may provide the most suitable method for determining the ordinary high water mark.

Further, use of hydrological data is specifically implied in some jurisdictions such as in Louisiana, where LA Civ. Code Articles 450 and 456 define the boundary between public trust ownership and riparian upland in navigable rivers as the "ordinary low stage of water," with the banks subject to private ownership although also subject to a public easement.

Therefore, water level records are potentially valuable evidence for determining nontidal water boundaries as they are with tidal water boundaries. Yet, resolution of the best method for interpretation of such data is still an unsettled question. The *Manual of Surveying Instruction* provides only very cryptic instruction on this topic:

> Practically all inland water bodies pass through an annual cycle of changes, between the extremes of which will be found mean (ordinary) high water.

Some case law also hints at possible approaches such as the following:

> This word (ordinary high water) … does not mean the abnormally low level of a lake during one of a series of excessively dry years, or the abnormally high level of a lake during an exceptional wet year or series of wet years, but the average or mean level obtaining (sic) under fairly normal or average weather conditions, allowing the proper range between high and low water mark in average years. (*Tilden v. Smith**)

Various approaches have been suggested over the years for the use of water levels to support or at least resolve ambiguity in physical evidence. One commonly suggested approach has been the use of the level representing a certain percentage of inundation. Such a method has not met judicial acceptance due to the arbitrariness of selecting a specific percentage of inundation. Furthermore, evaluation of this method suggests that, depending on the water body, a wide range of percentage values may be found for the ordinary high water line as found by more traditional physical indicators. Therefore, there is currently no widely accepted approach to the use of water level data.

Nevertheless, there is a recently proposed approach to the use of hydrological data that is ideal for use as collaborative evidence to resolve ambiguities between competing physical evidence and may also serve as the best evidence for the location of the boundaries if physical evidence has been obliterated. This approach is especially helpful for the location of the ordinary low water mark, where physical evidence may be hidden or obliterated by higher water levels. It is based on the frequently cited judicial definition for a boundary line "…where the presence and action of waters are so common and usual and so long continued in all ordinary years…"

Under this approach, daily water level readings over a long period of time are sorted by elevation. Then, the number of observations occurring in each

* Fla, 113 So 708 (1927)

range of elevations is plotted in a histogram. That process graphically reveals the range of elevations where the presence of the water has been for the longest periods of time. Invariably, this process reveals two "nodes," where the presence of the water has occurred for longer periods. Therefore, this process does not involve statistical averaging but instead is a method for determining the elevations at which the presence of the water is most common.

Illustration of typical evidence for a nontidal water boundary—The following will illustrate typical evidence used in a nontidal water boundary survey on a typical river.

Considering botanical evidence, the prominent line just below the base of the tree on the bank would be considered to be the ordinary high water line since it represents the lower limit of terrestrial vegetation, the upland limit of where vegetation has been wrested away by wave action or stream currents. Considering geomorphological evidence, the prominent escarpment at that same location provides additional evidence of the ordinary high water mark. Both of those features are at an elevation of approximately 89 feet with National Geodetic Vertical Datum of 1929 (NGVD29). Although not clearly visible in Figure 6.6, there is also a faint escarpment at the edge of the water at approximately 86 feet elevation, which would suggest the location of the ordinary low water mark.

Considering hydrological evidence, a nearby U.S. Geological Survey (USGS) gauge has recorded daily water level records for over 20 years (Figure 6.7). Using the method described previously for an approach for use of hydrographical data, the daily elevations were sorted by elevation and a count made of the number of days where the water level was in each foot of elevation during the period of record. Those counts were then used to create a

FIGURE 6.6
Typical North Florida river bank (photograph taken at very low water level to expose bank).

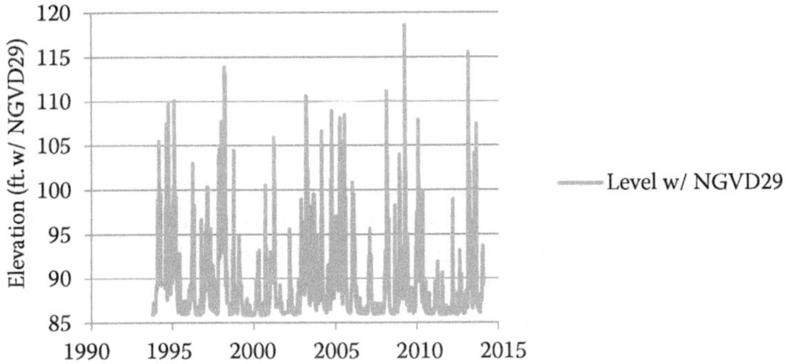

FIGURE 6.7
Daily water level records for North Florida River.

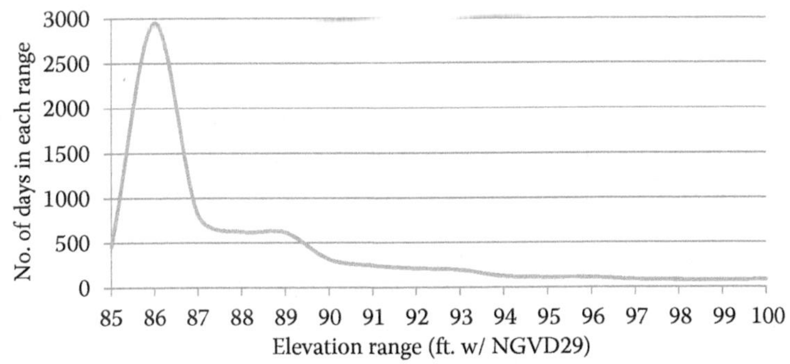

FIGURE 6.8
Histogram of daily water level records for North Florida River.

histogram (Figure 6.8). Unlike many histograms that use columns, a linear graph has been used for better resolution.

As may be seen, the histogram graphically depicts two "nodes" or elevation brackets where the "presence and action of the water" is most prevalent. The higher of these at elevation 89 feet would suggest the elevation for the ordinary high water mark. The lower, at about 86 feet, would suggest the elevation of the ordinary low water mark. Both of these elevations agree with the elevations of physical evidence on the ground. Notable, these agree with the physical features depicted in the photograph in Figure 6.6.

6.3.4 Boundaries Related to Nonpublic Trust Waters

In addition to the public/private boundaries discussed in the previous two sections, other frequently encountered water boundaries are those where a water body is the boundary line between two ownerships. Perhaps, the most elementary example of this type of boundary is that involving a stream as the boundary between two parcels of land. In such a case, when the deeds of either side call "to the stream," the center of the stream would be the boundary. There are two complicating issues relating to stream boundaries where a divergence among court opinions exists. These issues deal with the questions of what constitutes the "center" of the stream and how the lateral boundaries between the upland boundaries and the center of the stream should be run.

Regarding the center of the stream, one approach defines the boundary as the "thread of the stream," which is defined as the line lying equidistant between the banks. For example, one source, supported by a number of court cases, states as follows (Bade 1940):

> The term 'thread of the stream' means the geographic center of the stream at ordinary or medium stage of the water, disregarding slight and exceptional irregularities in the banks. It is fixed without regard to the main channel of the stream.... If the stream is made a boundary in a private conveyance, then the thread of the stream will be the stream boundary.

The second approach regarding the center of the stream defines the boundary as the *thalweg*, or the deepest part of the channel. As an example of this approach, one opinion (*Stubblefield v. Osborn**) states as follows:

> Upon principle, therefore, it would appear that the thread of a nonnavigable river is the line of water at its lowest stage. The thread or center of a channel, as the term is above employed, must be the line which would give to the landowners on either side access to the water, whatever its stage might be and particularly at its lowest stage.

When nonnavigable lakes or ponds form the boundary line between two or more parcels, approaches for location of the boundary line vary with the nature of the lake. For round lakes, the general rule is that a center point is selected at the geographical center (or arguably the deepest point in the lake) of the lake and partition lines are then run from the ends of the upland boundaries to the center point, forming "pie slices." For long or irregular lakes, the lines are typically run perpendicular to a center line.

6.4 Interpretation of Land Descriptions

Retracement surveys are often considered to consist of two distinct components. The first is determining the location of the boundary and the second is accurately measuring the boundary. Often, the first task is the most difficult

* Neb, 31 NW 2d 547 (1948)

and is also the most important. A venerable surveying text, *Boundaries and Landmarks* (Mulford 1912), expresses this sentiment succinctly as follows:

> ...it is far more important to have a somewhat faulty measurement of the spot where the line truly exists than it is to have an extremely accurate measurement of the place where the line does not exist at all.

6.4.1 Interpretation of Description Calls

A major part of determining the correct location of a boundary line is the correct interpretation of the land description for the tract being surveyed. Case law over the years has resulted in a body of guidelines for the best interpretation of descriptions. Some of the most salient are as follows:

- Distances: Unless otherwise stated, called distances are to be measured in a straight line along the shortest measurable distance and are horizontal distances.
- Calls using the word *to*: When calls use the word *to*, such as "to a concrete monument" or "to a river" or "to an adjoining property boundary," the actual direction and distance to the object called takes priority over any direction and distance cited in the description.
- Indeterminate fractional or area calls: Indeterminate fractional or area calls are those where the direction of the dividing line is not given or implied. Some of the principles that have been developed for such situations are as follows:
    - A called area of land on the side of a tract, such as "the northern 80 acres," should be interpreted as including the area in the form of a parallelogram, with the dividing line parallel to the side of the tract on which the area is located.
    - A called area of land in a specified corner of a tract should be interpreted as being a corner quadrangle with equal sides.
    - When the easterly and westerly sides of a lot are nearly parallel and either the easterly or westerly halves are described, the dividing line should be considered to be on a mean bearing between the two sides.
    - When the easterly and westerly sides of a lot are not parallel and either the easterly or westerly halves are described, the dividing line should be considered to be north–south.
    - When a description calls for a lot except for a specified half (such as except for the eastern half), the dividing line should be parallel to the boundary of the lot on the side of the specified half.
- Directions with or without "-ly": When compass directions (north, southeast, etc.) are used, they generally imply an actual bearing.

When compass directions with the suffix "ly" (northerly, southeasterly, etc.) are used, they generally imply relative position.

- Calls using the term *parallel*: Where straight lines are involved, the term *parallel* implies that the lines have the same bearing. For multicourse lines, each segment must be parallel. Curves are parallel if they are concentric with a common radius point.
- Calls using the term *tangent*: A *tangent* is the straight line adjoining a curve at either end. A tangent curve is one where the radius of the curve at the point of curvature is perpendicular to the straight line segment connected to the curve at that point. A *tangent of a curve* is the extension of a tangent from one end of the curve to its intersection, with the extension of the tangent from the other end of the curve. Compound or reverse curves are considered to be tangent if the radii are common or one is an extension of the other.
- Calls for fractional parts of public land sections: Calls for fractional parts of public land survey sections (SE 1/4, SW 1/4 of the NW 1/4, etc.) imply division of the sections by procedures prescribed in the *Manual of Surveying Instruction* (Bureau of Land Management 2009).
- Senior and junior rights: Where descriptions for adjoining parcels result in overlaps, the holder of the first grant or patent (the senior awardee) retains ownership of the overlap area, and the holder of the latter grant or patent (the junior awardee) loses the area. A common expression explains these rights: "The first deed is the best deed."
- Water boundary calls: When calls are "to" or "along" a water body, the location of the boundary will vary with the nature of the water body and prevailing state law. Generalized guidelines are as follows:
  - In navigable tidally affected waters, the call "to" or "along" a water body in most states should be interpreted as to or along the mean high water line, with the water below that line being considered as public trust waters. That line is the intersection of the rising shoreline, with the average elevation all the high tides over a complete tidal epoch of 19 years. There are exceptions to this principle. In several states, the line should be at the elevation of mean low water or at a specific distance seaward of the mean high water line. In other states and in Puerto Rico, where title to land has its origin in Spanish or Mexican sovereign grants, the line is based on the Roman Common Law, which holds that the correct line is "as far as the waves go at furthest" (Sandars 1874).
  - In navigable waters not affected by the tide, the call "to" or "along" a water body in some states should be interpreted as to or along the ordinary high water line. That line is where the constant presence and action of the water have marked upon the

soil a character distinct from that of the banks. In states that recognize the ordinary low water line, it would be along that line. In states that do not consider the beds of nontidal waters to be publicly owned, nontidal waters should be treated the same as nonnavigable waters.

- In nonnavigable waters, the call "to" or "along" a water body should be interpreted as to or along the center or thalweg of the water body. It should be noted that some states consider all waters affected by the tide, regardless of their navigability-in-fact, to be waters of the public trust. Thus, in those states, the correct line, in tidally affected waters, is the mean high water line for tidal waters regardless of navigability-in-fact.

6.4.2 Rules of Construction

When there is ambiguity in a description, "Rules of Construction" have evolved in various judicial opinions for guidance in resolving the ambiguity. Such rules of construction apply only when ambiguity exists in a description, and courts apply such rules only when a writing is subject to more than one meaning. When descriptions are not ambiguous, they must be followed as written.* In other words, in construing a writing, it is not permissible to interpret that which has no need of interpretation.

Unlike a settled rule of property, which has become a rule of law, rules of construction are subordinate and always yield to the intention of the parties, particularly the intention of the grantor, where such intention can be ascertained. Since all rules of construction are essentially only methods of reasoning that experience has taught are best calculated to lead to the intention of the parties, generally, no rule will be adopted that tends to defeat that intention.†

There is an abstract distinction between "construction" and "interpretation," in that "construction" is the drawing of conclusions from elements known from, given in, and indicated by the language used, while "interpretation" is the art of finding the true sense of the language itself or of any form of words or symbols.

In other words, "interpretation" is used with respect to language or symbols themselves, while "construction" is used to determine not the sense of the words or symbols but the legal meaning of the entire contract. However, in many contract cases, it would be difficult, as well as unrewarding, to determine where "interpretation" ends and "construction" takes over. It is

---

* *Gaweylak v. Cowie,* 86 N.W.2d 809, 350 Mich. 679 (1957); *Alabama Medicaid Agency v. Wade,* 494 So.2d 654, Ala. Civ. App. (1986), stating: "rules of construction should not be used in construing a deed if intention of parties to the deed, especially the grantor, can be ascertained form the document itself."

† *Re Estate of Fleck,* 154 N.W.2d 865, 261 Iowa 434 (1967); *Lassiter v. Goldblatt Bros., Inc.,* 41 N.E.2d 803, 229 Ind. 215 (1942).

doubtful if the abstract distinction between the terms is of any vital significance, and certainly, in common and legal usage, "construction" and "interpretation" are used interchangeably.*

The construction of deeds, as well as other legal documents, presents a question of law for a court to decide. It is the court's duty to interpret a deed in the light of the law in existence at the time of its execution and delivery, which must be read into and becomes an enforceable part thereof.†

Rules of construction include, but are not limited to, the following:

- The intent of the parties is the controlling consideration.‡
- In ascertaining the intention of the parties, separate deeds or instruments executed at the same time and in relation to the same subject matter, between the same parties, may be taken together and construed as one instrument.
- A deed must be construed as a whole, and a meaning given to every part thereof.
- Words in a deed are presumed to have a purpose.
- Documents are to be viewed in light of the surrounding circumstances.
- Descriptions are to be construed according to their plain terms, and words given ordinary meaning.
- A document should be construed in light of the law in existence at the time it was executed.
- A deed should be construed with reference to the actual state of the land at the time of its execution.
- Technical words should be construed according to their technical meaning.
- Words that have definite legal significance must be given their legal effect.
- A specific, or particular, description controls a general one.
- References are part of the description with as much effect as if copied into the document.
- Descriptions are to be construed against the grantor and in favor of the grantee.§
- A false or erroneous description, or statement, may be disregarded.

* See 17 Am Jur2d, Contracts, § 240. 30 Am Jur 2d Evidence, § 1065.

† See 17 Am Jur2d, Contracts, § 240. 30 Am Jur 2d Evidence, § 1065.

‡ Intent of the parties is to be gathered from the writing. *Smart v. Huckins*, 82 N.H. 342 (1926): the process of construction builds upon the language to develop the intention, and not the intention to develop the language. It is an application of the deed to the land, and not of the land to the deed.

§ A rule of last resort, when other rules fail. *Smart v. Huckins*, 82 N.H. 342 (1926).

- A meaning and intending clause will not limit or enlarge the grant.
- Punctuation is ordinarily given slight consideration in construing a deed.
- Words will control punctuation marks rather than the other way around; however, punctuation will be resorted to in order to settle the meaning of an instrument, after all other means fail.
- Courts generally take judicial notice of commonly used initials and abbreviations in descriptions.

Recommended Additional Reading on Land Boundary Surveying

Bureau of Land Management (2009). *Manual of Surveying Instruction*. Washington, DC: U.S. Government Printing Office.

Cole, George M. (1997). *Water Boundaries*. New York: John Wiley & Sons.

Robillard, Walter G. and Donald A. Wilson (2014). *Brown's Boundary Control and Legal Principles*. 7th Edition. New York: John Wiley & Sons.

Wilson, Donald A. (2008). *Forensic Procedures for Boundary and Title Investigation*. Hoboken: John Wiley & Sons, Inc.

Wilson, Donald A. (2015). *Interpreting Land Records*. 2nd Edition. Hoboken: John Wiley & Sons, Inc.

7

Cadastral Systems

Cadastres are the basic source of information for all aspects of land tenure and use, including the resolution of title and boundary disputes. In most countries, including the United States, extensive legal systems have developed around the original administrative systems and the records and information contained within those systems. In addition, cadastres have an important social function in serving as an archive that protects communities from commercial chaos following disasters such as floods, earthquakes, fire, hurricanes, or financial panics.

7.1 History of Cadastral Systems

A *cadastre* (also known as *cadaster*) is usually defined as a comprehensive register of the real property of a country, or a designated area. It generally employs the use of a cadastral map, or cadastral survey. It also generally includes details of ownership, tenure, precise location sometimes using geographical coordinates, parcel dimensions and area, any cultivations of rural parcels, and parcel values. Cadastres are traditionally classified as either juridical or fiscal cadastres. Juridical cadastres serve as repositories or registries of ownership and other land rights. These include land records recordation systems, which serve as simple public repositories of various land records, and more complex land registration systems, which record and also warrant title. As opposed to juridical cadastres, which are for the purpose of protection of land rights, fiscal cadastres serve as the basis for the comprehensive, equitable, and accurate taxation of land (Larsson 1991).

The word *cadastre* was adopted into the English language through French, from the Late Latin *capitastrum,* which signifies a register of the poll tax, and the Greek *katastikhon,* which is a list or register, from the Greek *kata stikhon.* The latter literally translates to "down the line," in the sense of "line by line" along the directions and distances between property corners designated and marked by monuments in the metes and bounds of the tract of land.

As the intensity of the use of land increased in early society, it became increasingly important to protect investments in that use. As a result, that need for protection led to public registries of land ownership to allow legal protection of land claims very early in society's evolution. Thus, juridical cadastres were probably the first type of cadastre developed. In some primitive cultures, transfers of land were required to take place in the presence of the chief and elders. Later, it became standard practice in more developed cultures to deposit and officially register deeds of sale at a court or with a notary public (Larsson 1991). As evidence of this, there are biblical references to the public filing in such systems of deeds for land transactions. Similar procedures were used early on in Assyrian-Babylonian and Egypt, where land transactions had to be witnessed and deeds filed in a royal repository as early as 3000 BC (Larsson 1991). Although few details are available, evidence suggests that the registry was based on field surveys of land parcels.

Fiscal cadastres for taxation purposes also developed in early civilizations. For example, at the end of the third century AD, the Romans surveyed the territories that they occupied and recorded them for taxation purposes. In China, a taxation system supported by land survey records was in existence around AD 700. In India, Raja the Great ordered a revenue survey around AD 1000, and in Sweden, King Gustav I ordered a survey of all taxable lands in 1540.

Possibly the most renowned of the early fiscal cadastres was the Domesday Book established in England in 1086. That cadastre was created from a census of all of England ordered by William the Conqueror after he defeated the Saxons. The resulting product involved a record for each land parcel that included identification of the landowner, area, a description of land use for the parcel, number of tenants, and quantity and type of livestock. The Domesday Book did not involve a map but included only tabulated information. The information in that book was frequently used in litigation in the Middle Ages. Even today, it is used in certain cases.

Another frequently cited early cadastre was developed under the orders of Napoleon in France in 1807. That system included a map created from a survey based on control points determined by triangulation. Unique parcel numbers were developed for each land parcel in the country, and the ownership, land area, land use, and value were recorded for each. The French cadastre served as a model for central Europe and influenced the development of similar systems in a number of other countries in that area. Although that cadastre was developed primarily for the purposes of tax assessment, it was recognized by Napoleon himself as also having significant benefits for security of land title, as evidenced in his writings from his exile on the island of St. Helena as follows (Larsson 1991):

> The cadastre just by itself could have been regarded as the real beginning of the Empire, for it meant a secure guarantee of land ownership, providing for every citizen certainty of independence.

In 1785, the United States began the large cadastre known as the U.S. Public Land Survey to survey, subdivide, and spatially identify the public domain lands into sellable parcels and to provide for the orderly sale and settlement of publicly owned land. Sometimes referred to as the rectangular survey system, the system is commonly known today as the PLSS. This system was designed to survey the land into rectangular lots before sale to prevent overlapping claims, gaps, or gores; to provide for the buyer to take poor land along with the good; and to simplify deeds and registration. That system subdivided the land into six-mile square townships with boundaries in the cardinal directions and further subdivided each township into 36 one-mile square sections. It was the first mathematically designed and nationally conducted cadastral system in any modern country. The system was designed primarily by a committee headed by Thomas Jefferson. Having a background in land surveying and law, he understood the operation and value of an orderly system of land and parcel recognition. After its beginning at a point at the intersection of the western Pennsylvania boundary and the northwestern bank of the Ohio River (Figure 7.1), the basic framework of the survey has remained essentially unchanged and now serves as the basis for land title in a large portion of the country. (See Section 4.4 for a complete discussion of this system.)

In Canada, a cadastre known as the Dominion Land Survey was begun in 1871 to divide most of Western Canada into one-square-mile sections for agricultural and other purposes. It is based on the layout of the U.S. Public Land Survey System, although with some differences. As with the U.S. system, it divided the land into townships of roughly six miles by six miles, which were further divided into sections of one mile by one mile. In addition to differences between the Dominion Land Survey and the U.S. Public Land Survey in how the sections are numbered, many of the townships in the Dominion Land Survey have "road allowances" between certain sections of the township. These are strips reserved for roadways running along outside the perimeter of sections. The width and spacing of the strips vary with location and instructions prevailing at the time of survey. As an example, surveys performed after 1881 have one-chain-wide road allowances every mile going east–west and every two miles going north–south. The Dominion Land Survey is the dominant land survey in all of the prairie provinces (Alberta, Saskatchewan, and Manitoba) and is also used in portions of British Columbia. Currently, over 310,000 square miles has been covered.

Thus, cadastral systems have a long history of application. Such systems have existed almost from the beginning of formal land ownership. The development of cadastral systems was closely linked to the intensive use of land and the development of private property rights. With private rights to land came the need of the government to manage those rights and provide protection to the landholders for those rights and also a parallel need to obtain appropriate taxes from land owners. Those needs precipitated the development of cadastral systems.

Although many cadastres and land registries were originally developed for the purposes of taxation, alternate uses developed from even the earliest land registries. Today, land registries serve an essential role in ensuring

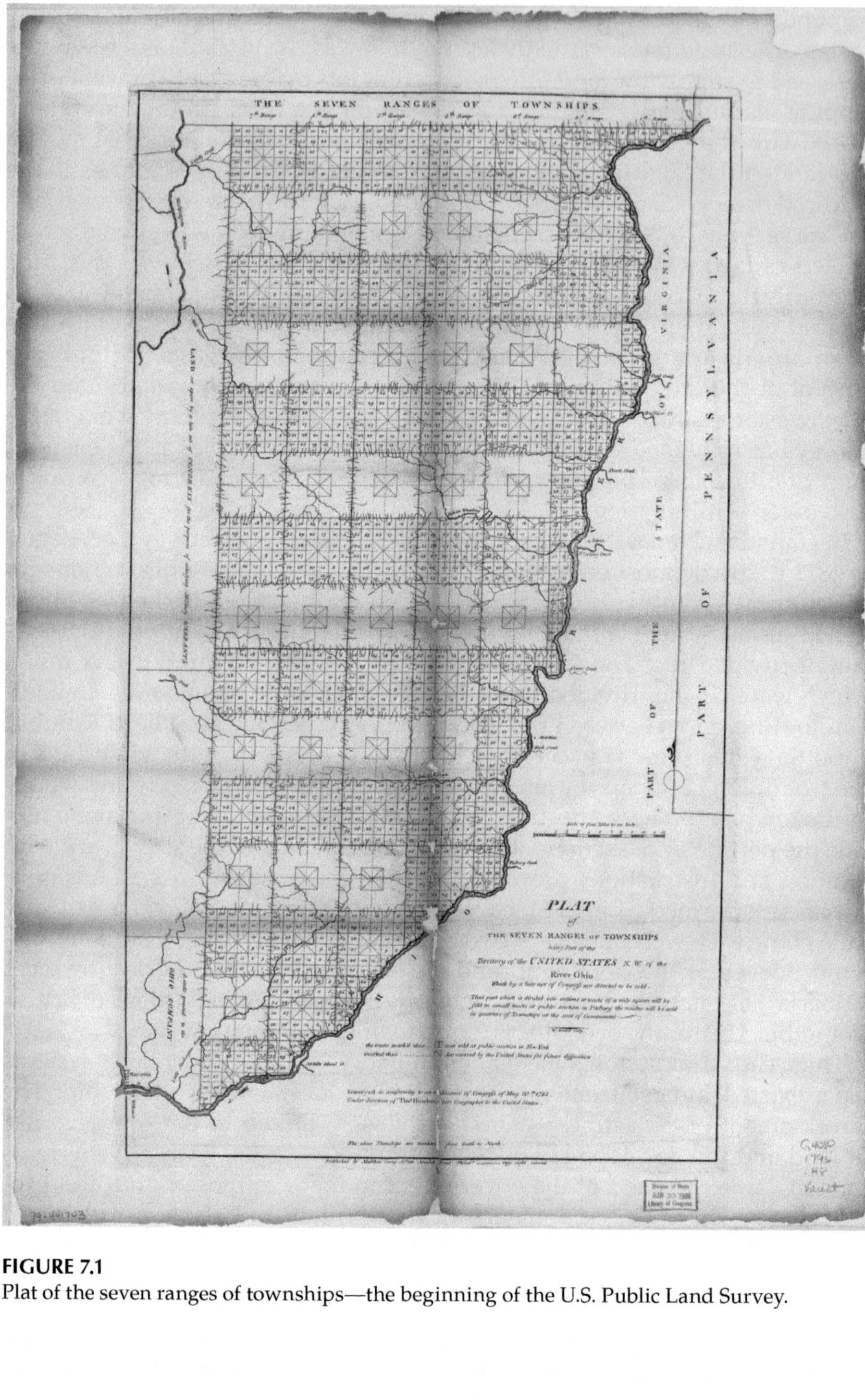

FIGURE 7.1
Plat of the seven ranges of townships—the beginning of the U.S. Public Land Survey.

security of title to land, land planning and regulation, emergency planning, real estate and financial services, transportation and utility planning, school zoning, and various other societal functions.

7.2 Juridical Cadastres

7.2.1 Land Registration Systems

Internationally, *land registration systems* are the most common type of juridical cadastres. Such systems serve two functions. They allow interested parties to see who owns what, and in addition, they provide a government-based registration or certification of ownership to each land parcel. Such cadastres are primarily for the purpose of security of title and not for apportionment of taxes.

The most widely used method for land registration is called the Torrens land registration system. The Torrens system is named after its creator, Sir Robert Torrens. A boom in land speculation and a haphazard grant system resulted in the loss of over 75% of the 40,000 land grants issued in the colony (now state) of South Australia in the early 1800s. To resolve the deficiencies of the common law and deeds registration system, Sir Robert, then a member of the colony's House of Assembly, proposed a new title system in 1858, which was quickly adopted.

Although the system bears his name, Torrens acknowledged that he adapted his proposals from earlier systems of transfer and registration. The practice of title registration had been customary in parts of Europe prior to the 14th century. When Torrens began to research the subject, he found that similar systems had existed in Austria, Hungary, and Bavaria for centuries. Nevertheless, he did not copy them but devised a new one based on the system in use for registering merchant ships in the United Kingdom. That ship registry assigned each ship owner a certificate that included certain information. When a vessel was sold, the seller surrendered the certificate for cancellation and a new certificate was given to the new owner.

Under the Torrens system, a register of land holdings is maintained by the state, which guarantees title to those included in the register. Land ownership is transferred through registration of title rather than by use of deeds. Its main purpose is to simplify land transactions and to certify to the ownership of an absolute title to realty.

The Torrens system is similar to a judicial process. The registration process is initiated by a petition. Following public notice and a title search and adjudication by the title registration office, a final decree of ownership and title is issued. The adjudication process is obviously a key element in the registration process. It involves an authoritative determination of all existing rights in a particular parcel of land. To facilitate the adjudication process, it is important that the cadaster include ownership information for all land units in the covered area. As a result, unlike less complex recordation systems, land registration systems need a map element to ensure complete coverage as well as to illustrate the spatial relationship among parcels.

The requirement for a cadastral map for land registration systems as well as for a fiscal cadaster maintained for land taxation purposes has resulted in conflicts in some jurisdictions. It is obviously critical that maps used for the two separate purposes agree. Ideally, the same map and supporting data should be used. Yet, the land registration and tax assessment processes typically involve different government entities, making coordination a sometimes difficult task.

After its development in Australia, the Torrens system was adopted by a number of other British colonies as well as other nations. The system was extended into various Canadian provinces between 1870 and 1885.

In the past, this system was also the subject of considerable interest in the United States. Torrens acts were adopted in 20 states and territories between 1895 and 1917, but only 11 states now have such a system in effect and the system is used to a substantial extent today in only 3 states: Hawaii, Massachusetts, and Minnesota. In Hawaii and Massachusetts, Torrens is used statewide. In the other states, use has been limited to a few localities, such as in Illinois, where it is used only in Cook County. In no state or locality are a majority of parcels registered under Torrens. The highest incidence of use is in Hawaii, where nearly 45 percent of all parcels are registered.

In America, Illinois was the first state to adopt a Torrens Title Act. The Chicago Fire of 1871 had destroyed Cook County real estate records. Since Chicago, and Cook County in general, was rapidly expanding, providing an adequate verification of title was difficult. The destruction of the records by the great fire had given the Chicago Title and Trust Company, which possessed the only set of abstracts extant, a practical monopoly of the title business, and owners were largely at the mercy of the title company. In response, the Illinois legislature passed the first Torrens Title Act in the United States in 1895, although it was declared unconstitutional by the Illinois Supreme Court. In 1897, the legislature passed an amended version, which became the first Torrens Title Act in the United States. That law was upheld by its Supreme Court and was adopted only in Cook County. By 1937, approximately 20% of the 1,300,000 parcels of land in Cook County had been registered under the system. Due to declining use, in January 1992, the Illinois legislature began the process of phasing it out.

The Commonwealth of Massachusetts was next with a Torrens statute in 1898, which was later upheld by her Supreme Court. Alfred Hemingway, of the Boston Bar, was appointed as a one-man commission to draft the law. The result of his work was the Massachusetts registration law, which has become, in many respects, the model law in America; the Massachusetts Act assigns all the duties of the Torrens law to a separate land court.

California passed a Torrens law in 1897. Yet, since it failed to provide an assurance fund, titles under that system did not become absolute until five years had elapsed. As a result, the law failed, and a new law was adopted in 1914 amending the 1897 law. Confidence in the system was eroded in 1937, when the entire state assurance fund was wiped out by a single claim. California essentially abandoned the system in 1955 and repealed the law in 1958.

A Torrens law was passed in Minnesota in 1901 for counties with a population of over 75,000 and was amended in 1909 to make the law applicable to the entire state. Torrens is still in existence in Minnesota today, and Hennepin County, the most populous county in the state, has more registered property than any of the other counties.

Oregon also passed its law in 1901, which was abolished in 1972. Colorado followed in 1903. Also in 1903, Hawaii, as a territory of the United States, instituted its Torrens law.

In 1907, Washington passed its law. The majority of registrations have been in Seattle and surrounding neighborhoods.

In 1908, New York enacted its Torrens statute, and its use was primarily around Buffalo and on Long Island. The law, as amended, went into effect in 1910. The system was abolished by the New York Legislature as of January 1, 2000.

A Torrens law was passed in Ohio in 1896, although it was held as invalid. Yet, in 1912, a new statute was authorized and enacted by its new Constitution. That law was mandatory at first but was made optional in 1915. A recent survey of the 88 Ohio counties demonstrated that 48 counties have never had registered land. Of the counties that have had Torrens parcels, 26 counties have chosen to abolish the system, including Butler County in 2014.

In North Carolina, large tracts of swamp land purchased by lumber companies have been registered under the Torrens system. That was for the purpose of defining boundaries and attaining a legal defense to the claims of squatters.

Mississippi enacted Torrens legislation in 1914, and Nebraska, in 1915. Also in 1915, an amendment was adopted to the Constitution of Pennsylvania authorizing legislation providing for a system of registering and insuring land titles by the state or counties thereof.

South Carolina instituted a law in 1916. The State of Virginia amended her constitution in 1916 in order to adopt the system, being the first state to provide in its constitution for the enactment of a Torrens Act. Five other states (Georgia, Utah, North Dakota, South Dakota, and Tennessee) passed their laws in 1917.

The State of Iowa uses a form of Torrens title system. That state does not permit private title insurance and instead operates a guaranty fund that insures property titles, essentially identical to private title insurers. In 1947, Iowa banned title insurance after a series of insurance bankruptcies in Sioux City and created a statewide title guaranty program as an alternative. In

1985, Iowa developed Title Guaranty, the state-sponsored title guaranty, to ensure that mortgages could be sold on the secondary market.

The U.S. Government also became involved in the Torrens system during the early 1900s. This included introducing it into the Philippines (1902), Hawaii (1903), and Puerto Rico.

As may be seen from its history, the Torrens system was at one time authorized in numerous parts of the United States. Yet, it generally has not been accepted and is rarely used in the jurisdictions in which it is still authorized. Various reasons have been advanced for that rejection. There are obviously significant advantages to a government-certified title to land offered by the system. It provides assurance to the registered owner that he or she is the rightful owner to the titled land. It also precludes the loss of land by adverse possession or prescriptive use. Nevertheless, there are also disadvantages to the use of such a system. Certain claims, including tax liens, interests arising from bankruptcy procedures, and mechanic's liens, may override the Torrens certification. In addition, the registration process can be both lengthy and expensive to both users and the public, which is certainly one reason for the nonuse of the system. Yet, another reason is the strong, functioning land recordation system that exists in the United States (see Section 7.2.2). The use of that traditional system fits well with the general political philosophy in the United States in that costs for a title search are borne by the individual needing that service rather than the general public having to support an expensive cadastral system. Further, the traditional land recordation system meshes well with a general mistrust of government agencies performing functions, such as title searches, that may be adequately accomplished by the private sector.

7.2.2 Land Records Recordation Systems

Within almost all of the United States, rather than a land registration system, a more basic type of juridical system, a land records recordation system, is utilized. Such a system is basically a filing system for records dealing with title to land. Such records include deeds, easements, mortgages, liens, leases, and similar documents. Typically, formal subdivision plats are also recorded in these systems as are probate records for wills and other judicial actions affecting land ownership. As with land registration systems, land records recording systems are primarily for the purpose of providing security of title rather than apportionment of taxes.

Land recordation systems provide public notice of ownership, leaving evaluation of the validity of that claim up to the individual land owner on a need basis. The function of such systems is to make available to the public all information concerning interests in real estate. Land recordation systems, as typically structured in the United States, do not guarantee validity of title. Rather, by making deeds and other documents affecting land title publicly available, they provide constructive notice of ownership. By providing a database of land title records, such a system allows for searches to examine and determine the validity of any competing claims.

The effect of recordation is to give the recorded instrument a priority over all instruments that are either unrecorded or not recorded until later. The basic difference between deed recordation and the Torrens system is that the former involves recordation (or registration) of evidence of title while the latter involves registration of actual title.

The recordation of deeds in a public office is an American innovation, reportedly first practiced in the Plymouth, Massachusetts, colony about 1624 (Shick and Plotkin 1978). In 1636, the Plymouth Bay Colony had adopted its first recordation law, requiring that "all sales exchanges giftes mortgages leases or other Conveyances of houses and lands the sale to be acknowledged before the Governor or any one of the Assistants and committed to publick record." Similarly, in 1639, Connecticut insisted that "all bargaines or mortgages of land whatsoever shall be accounted of no value until they be recorded." By the time of the American Revolution, every English colony had adopted statutes requiring that parties to a land transaction, including conveyance of mortgages, must record their names and a description of the property in a public office designed for that purpose. After the American Revolution, with the acquisition of the Northwest Territories and the Louisiana Purchase, recording statutes were among the first laws to be legislated, replacing the French and Spanish systems. In almost all cases, land record offices were highly decentralized and are today typically maintained at the county or local level.

While recordation of a deed only serves as evidence of title, each recorded document regarding a parcel of land becomes part of a chain of title or history of all transactions affecting the title to the parcel. Thus, examination of such a chain of title allows identification of any competing claims and allows an interested party to determine the validity of title.

Over time, a land records recordation system can grow to contain thousands and even millions of documents. Therefore, the type of the indexing system maintained for such a system is important to facilitate title searches. Most land recordation systems in the United States traditionally used a grantor–grantee index. Such an indexing system involved two index files. Traditionally, the indices were maintained in two bound volumes, one for grantors and one for grantees. Separate pages were maintained in each volume for each letter of the alphabet. When an instrument was recorded, an entry was made describing the document in both indices on the appropriate page based on the first letter of the surname of both the grantor and grantee. A title search in a system with a grantor–grantee index consists of tracing the ownership history of a parcel by searching for each consecutive grantor or grantee on the appropriate pages.

An alternate type of index system used by some jurisdictions is a tract index. The traditional format for this type of index, also known as a geographic index, is a bound volume with separate pages for each land section, subdivision, or other tract. An entry describing the document is made on the appropriate page in the index when a document is recorded. With this type of index, a title search for a particular parcel can be limited to an examination of only those entries in one subsection of the index.

The objective for a standard title search for title purposes is the identification of transfers of ownership, conflicting titles, judgments against the land, liens for unpaid taxes or other liens against the property, and pledges of real property as security for loans. In addition, a great deal of information essential for boundary surveys may be obtained from a complete title search, including identification of junior–senior rights, the basis for any bearings used in the description to allow more precise retracement, easements that are not mentioned in the current deed, boundary agreements affecting the location of boundaries, transcription errors that may have been passed down the chain, and the true intent of the original parties when the subject tract was created (Robillard and Wilson 2011).

Regarding title searches, it should be recognized that a standard title search as described previously is only one part of a complete title search. It also should include a boundary survey (McEntyre 1978). A complete title search should include determination of the accuracy of the description as well as its location on the ground; examination of the description of adjacent parcels for boundary conflicts; and examination of the described land for evidence of possible prescriptive easements or adverse possession, encroachments, discrepancy between possession and the description, access problems, the existence of publicly owned waters, and changes due to erosion, accretion, and avulsive activity of the boundaries of such waters. Such factors obviously affect title, but can be determined only by a field survey.

Many states as well as some bar associations have drafted "Title Examination Standards," providing standard rules and procedures for title examinations based upon local law and experience in title matters. Such standards are intended to provide uniformity in the procedures used for title searches.

The title search process typically involves three steps:

- A chain of title is constructed of all previous owners of the parcel being researched. By searching for the name of the current owner in the grantee index, the name of the seller may be found. That new name is then use to identify the previous owner and the process continued. Ideally, the chain of title should be researched from the present back to the original conveyance from a sovereign.
- Each owner in the chain is then searched forward in time until the owner granted it to the next person in the chain. This step is to ensure that the previous owner did not first sell the land to other parties than those in your chain, thus creating a competing claim.

- The actual document for each transaction in the chain of title is then examined to ensure that it included the same land as that being researched and that each transaction was properly executed and did indeed represent an actual conveyance of title.

Most jurisdictions today have computerized their indexing system. This offers a method for indexing that provides for a highly efficient approach to title searches. Properly structured, a thorough title search may be rapidly conducted based on either the tract in which the parcel is located or on the grantors and grantees.

Examples of both grantee and tract indices may be seen on the Internet at the Bureau of Land Management's internet web site. On that site, the original conveyances out of the United States (for the public land states) may be searched by entering the section of interest or by entering the name of the grantee.

A land recording system does not provide protection against factors such as adverse possession, prescriptive easements, and errors in the land description, which can be discovered only by a physical survey of the land parcel. In addition, a land recordation system offers little protection against unrecorded or fraudulent instruments or other types of errors or fraud.

Many jurisdictions have developed marketable record title statutes that can substantially shorten the period over which a title search must be conducted. For example, the State of Florida has such a statute reading as follows (Section 712.02, Florida Statutes):

> **Marketable record title; suspension of applicability.**—Any person having the legal capacity to own land in this state, who, alone or together with her or his predecessors in title, has been vested with any estate in land of record for 30 years or more, shall have a marketable record title to such estate in said land, which shall be free and clear of all claims except the matters set forth as exceptions to marketability in § 712.03. A person shall have a marketable record title when the public records disclosed a record title transaction affecting the title to the land which has been of record for not less than 30 years purporting to create such estate either in:
>
> (1) The person claiming such estate; or
> (2) Some other person from whom, by one or more title transactions, such estate has passed to the person claiming such estate, with nothing appearing of record, in either case, purporting to divest such claimant of the estate claimed.

The effect of such an act is that it may be necessary to search back only 30 years to look for conflicting deeds and thus may eliminate having to search back to the original patent from a sovereign for a marketable title. Nevertheless, for obtaining maximum information for boundary survey purposes and for some title purposes, it is important to research the title back to the original patent. (See Section 6.2.1, Data Collection, for a discussion of the importance of this in boundary surveying.) The following cite from a Maryland case emphasizes this point.

> ...a requisite for valid title to real property is an original conveyance of public land from the State. Absent such a conveyance, one purporting to transfer an ownership interest in such property transfers nothing, and no quantity of successive transfers by deed nor the mere passage of time will metamorphose good title from void title. (*SkiRoundtop v. Wagerman**)

As an illustration of the continuing evolution of land recordation system in the United States, a discussion of the development of the public records system of Jefferson County, Florida, is presented: Jefferson is a predominately rural county in northern Florida stretching from the Gulf of Mexico to the Florida/Georgia border. The county was established in 1827, shortly after the Territory of Florida was acquired from Spain by the United States. The subdivision and sale of the public lands of the new territory had begun shortly before the county was established, and a land rush for the rich land in the area was in progress when the county was established (Shofner 1976).

The federal patents granted to settlers in the area were sold from the land office in nearby Tallahassee, so the original documents in the chain of title for land in the county are not contained in the county public records. However, there was an active resale land market ongoing in the region, so immediately after the formal establishment of the county, the first deeds in the county's public records were recorded. The first deed, representing the sale of land by James B. Watts in Section 30, Township 2 North, Range 5 East (Figure 7.2), was recorded by hand copying it into a large ledger on August 6, 1827. Interestingly, that land had been granted to Watts by a federal patent in June of the previous year.

The land recordation system for the county is maintained in the office of the County Clerk of Courts in the county court house. During the period from the first recorded deed in 1827 through 1934, there is no record of an index to the recorded documents being maintained. However, in 1934, a cumulative grantor/grantee index for the 1827–1934 period was prepared (Figure 7.3) as part of a program of the Works Progress Administration (Shofner 1976). Additional manually recorded grantor–grantee indices were maintained after that time until 2007. At that time, a switch was made to a computerized indexing system (K.B. Reams, Clerk of Courts, Jefferson County, personal communication, 2009).

Currently, in Jefferson County, as each document is presented for recording, an entry is made in a computerized indexing template requiring a selection of a type of instrument (e.g., deed, easement, mortgage) as well as

---

* 79 Md.App. 357, 556 A.2d 1144 (Md.App. 1989)

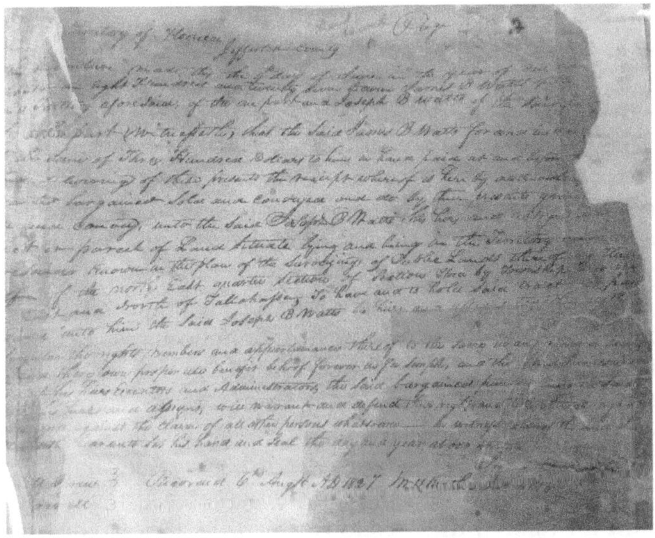

FIGURE 7.2
Photo of the first deed recorded in the public records of Jefferson County, Florida, August 9, 1826.

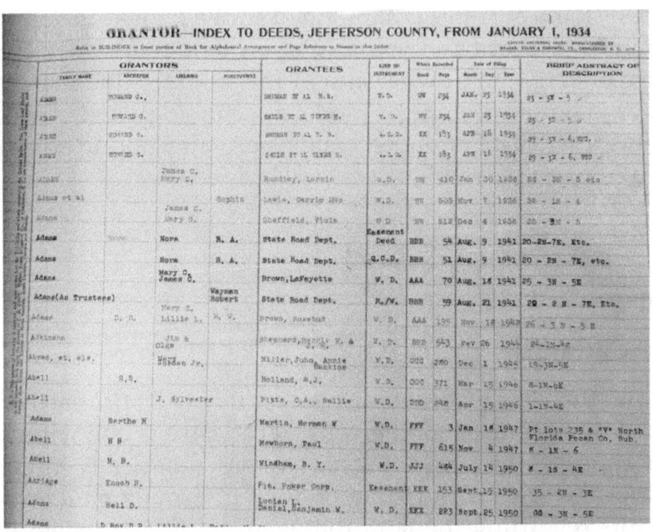

GRANTOR—INDEX TO DEEDS, JEFFERSON COUNTY, FROM JANUARY 1, 1934

GRANTORS Family Name				GRANTEES	Kind of Instrument	Where Recorded Book	Page	Date of Filing Month	Day	Year	BRIEF ABSTRACT OF DESCRIPTION
[illegible]	Howard C.			Shuman et al [illegible]	W.D.	WW	234	JAN.	23	1934	23 - 5N - 5
[illegible]	Howard C.			[illegible]	W.D.	WW	234	JAN	23	1934	23 - 5N - 5
[illegible]	Howard C.			Shuman et al [illegible]	[illegible]	XX	183	APR	16	1934	29 - 5N - 6, etc.
[illegible]	Howard C.			[illegible]	[illegible]	XX	183	APR	16	1934	29 - 5N - 6, etc.
Adams		James C. Mary C.		Bundley, Larkin	W.D.	[illegible]	410	Jan	30	1936	26 - 3N - 5 etc
Adams et al			Sophia	Lewis, Carrie Mae	W.D.	[illegible]	505	Nov	7	1936	32 - 1N - 4
Adams		James C. Mary C.		Sheffield, Viola	W.D.	[illegible]	512	Dec	4	1936	25 - 3N - 5
Adams	[illegible]	Nora	R. A.	State Road Dept.	Easement Deed	BBB	54	Aug.	9	1941	20-2N-7E, Etc.
Adams		Nora	R. A.	State Road Dept.	Q.C.D.	BBB	51	Aug.	9	1941	20 - 2N - 7E, etc.
Adams		Mary C. James C.		Brown, LaFeyette	W. D.	AAA	70	Aug.	18	1941	25 - 3N - 5E
Adams(As Trustees)			Wayman Robert	State Road Dept.	R./W.	BBB	59	Aug.	21	1941	29 - 2 N - 7E, Etc.
Adams	D. R.	Mary C. Lillie L.	R. V.	Brown, Rosebud	W. D.	AAA	195	Nov	18	1942	26 - 3 N - 5 E
Atkinson		Jim & Olga		Sheppard, [illegible]	W. D.	BBB	643	Feb	26	1944	24-1N-4E
Abrams, et. als.		Mary Jordan Jr.		Miller, John, Annie Rankins	W.D.	CCC	280	Dec	1	1944	15-3N-5E
Abell	R.B.			Holland, M.J.	W.D.	CCC	371	Mar	15	1946	8-1N-6E
Abell		J. Sylvester		Pitts, C.A., Sallie	W.D.	DDD	248	Apr	15	1946	1-1N-4E
Adams	Berthe H			Martin, Herman W	W.D.	FFF	3	Jan	18	1947	Pt lots 235 & "V" North Florida Pecan Co. Sub.
Abell	H B			Mewborn, Paul	W.D.	FFF	615	Nov	4	1947	8 - 1N - 6
Abell	H. B.			Windham, B. Y.	W.D.	JJJ	484	July	14	1950	8 - 1S - 4E
Abrige	Enoch H.			Fla. Power Corp.	Easement	KKK	153	Sept.	15	1950	35 - 2N - 3E
Adams	Bell D.			Lucian L. Daniel, Benjamin W.	W. D.	KKK	223	Sept.	25	1950	08 - 3N - 5E
Adams	D. Roy, D. R.	Lillie L.	[illegible]	[illegible]	[illegible]	[illegible]	[illegible]	[illegible]	[illegible]	[illegible]	[illegible]

FIGURE 7.3
Typical page of the Grantor Index for Jefferson County, Florida, for 1827–1834.

identification of the grantor and grantee, the description of land involved, information as to where a scan of the document is filed, and other administrative details. The document is then digitized and the original document, stamped with recordation information, is returned to the person recording it. Paper copies of documents recorded since 2007 are not maintained in the public records. A digital copy of the document is sent electronically to the County Property Appraiser, where the county's fiscal cadastre is maintained (see next section). That office uses the information from the document for changes to the county's fiscal cadastre as warranted. A digital copy of the document is also uploaded to a statewide web site maintained by the Florida Association of Court Clerks to allow online public access to the documents.

All documents recorded in 1973 and afterward are now maintained and may be viewed on the Florida Association of Court Clerks' web site (www.myfloridacounty.com). Documents recorded earlier have not been digitized since, based on Florida's Marketable Record Title Act, many title searches do not have to go back further than 30 years (yet, as previously mentioned, it is important for many purposes to trace a title back to the original patent from the sovereign). Users of the public record system in Jefferson County now have two options. They can elect to do traditional courthouse research, where the records, both paper and digital, are available for public scrutiny. Alternatively, they can log onto the previously mentioned web site, where they can search through all of the records recorded since 1973 using any of various search keys (grantor, grantee, location, type, date, record book, and page). Thus, a complete title search, as far back as 1973, can be performed online.

The system described here for Jefferson County is fairly typical of the land recordation systems currently maintained in the United States. While these systems contain volumes and volumes of ancient paper documents providing a history of early land ownership in this country, the more progressive systems have taken advantage of modern technology and moved to digital filing systems. Increasingly, those systems are also taking advantage of the Internet to allow ease of public access to land records.

7.3 Fiscal Cadastres

As opposed to the land recordation systems that exist primarily for the purpose of providing security of title, a fiscal cadastre is one with the primary purpose of apportioning taxes. Such a system is basically a file of all land parcels within the system's jurisdiction with information on ownership and various factors regarding the land that would affect its value. An essential component of most fiscal cadastres is an up-to-date property map. Thus, unlike the land records recordation system, a fiscal cadastre typically has a cartographic component. As with recordation systems, fiscal cadastres in the United States are typically decentralized and located in the office of the county tax assessor/property appraiser. In Puerto Rico, this type of system is maintained in regional offices of the Centro de Recaudación de Ingresos Municipales.

The typical fiscal cadastre has two major components: a property map (Figure 7.4) depicting the location and configuration of each parcel in the jurisdiction and an appraisal file (Figure 7.5) for each parcel. A property map is not a topographic map showing the physical features on the land. Rather, its main purpose is to depict the boundaries of parcels that are typically invisible to the eye on the ground. In addition to parcel boundaries, such maps may or may not also depict structures on the parcels. Often, this map is photo based. Such a map is designed to ensure that all land is accounted for in the taxation system and also to allow detection of overlapping parcels to prevent double taxation. Today, most property maps are in the form of a geographic information system (GIS). Such a system allows not only the depiction of each parcel in relation to other parcels but also the linking of data files to each parcel. Pertinent information from the appraisal file for each parcel is available with a click of a mouse on a parcel in the GIS system.

The appraisal file can contain a wide variety of information on the land parcel. In addition to ownership information, that file typically contains a condensed version of the land description for the parcel, a reference to the most current deed to the parcel, purchase price and date, information on

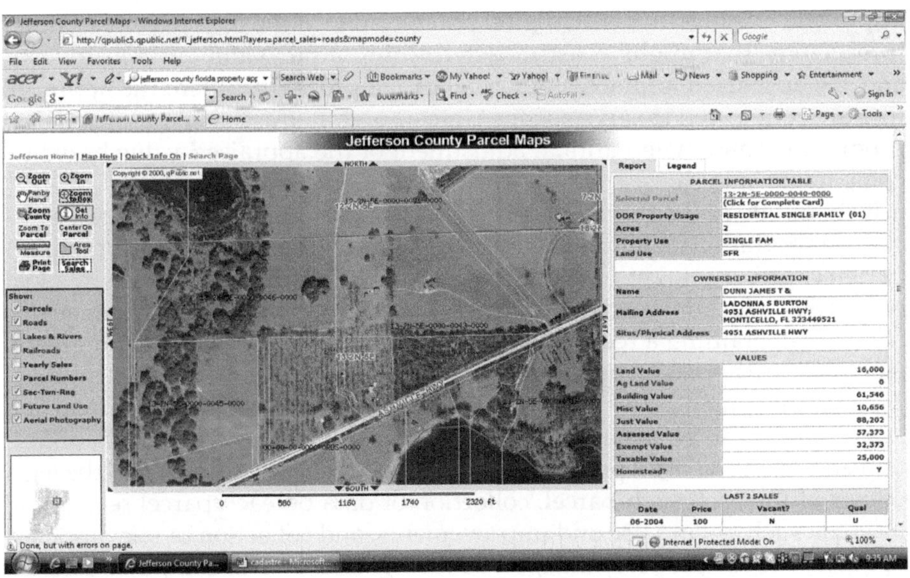

FIGURE 7.4
Typical screen in a property appraisal map in GIS format. (Screenshot used with permission of the Schneider Corporation.)

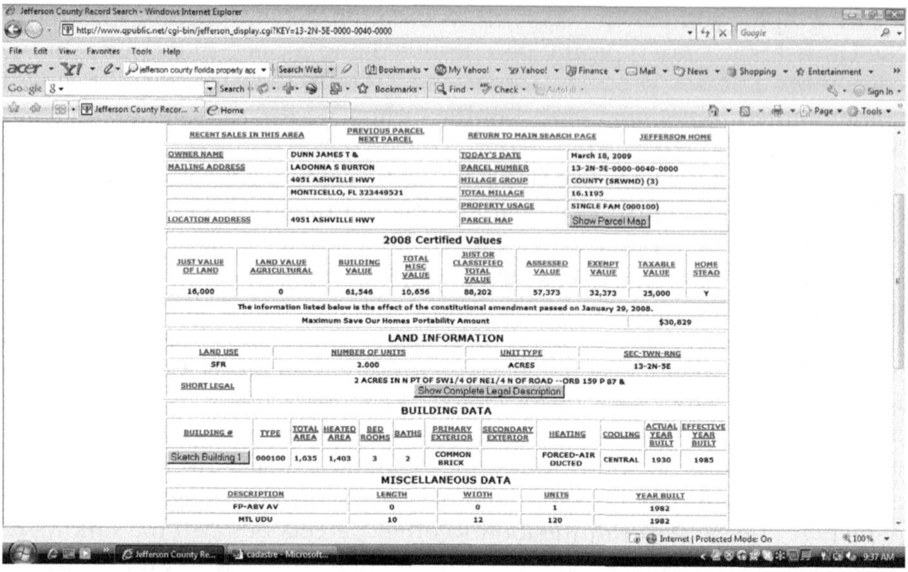

RECENT SALES IN THIS AREA | PREVIOUS PARCEL / NEXT PARCEL | RETURN TO MAIN SEARCH PAGE | JEFFERSON HOME

OWNER NAME	DUNN JAMES T &	TODAY'S DATE	March 18, 2009
MAILING ADDRESS	LADONNA S BURTON	PARCEL NUMBER	13-2N-5E-0000-0040-0000
	4951 ASHVILLE HWY	MILLAGE GROUP	COUNTY (SRWMD) (3)
	MONTICELLO, FL 323449521	TOTAL MILLAGE	16.1195
		PROPERTY USAGE	SINGLE FAM (000100)
LOCATION ADDRESS	4951 ASHVILLE HWY	PARCEL MAP	Show Parcel Map

2008 Certified Values

JUST VALUE OF LAND	LAND VALUE AGRICULTURAL	BUILDING VALUE	TOTAL MISC VALUE	JUST OR CLASSIFIED TOTAL VALUE	ASSESSED VALUE	EXEMPT VALUE	TAXABLE VALUE	HOME STEAD
16,000	0	61,546	10,656	88,202	57,373	32,373	25,000	Y

The information listed below is the effect of the constitutional amendment passed on January 29, 2008.

Maximum Save Our Homes Portability Amount $30,829

LAND INFORMATION

LAND USE	NUMBER OF UNITS	UNIT TYPE	SEC-TWN-RNG
SFR	2.000	ACRES	13-2N-5E

SHORT LEGAL: 2 ACRES IN N PT OF SW1/4 OF NE1/4 N OF ROAD --ORB 159 P 87 & [Show Complete Legal Description]

BUILDING DATA

BUILDING #	TYPE	TOTAL AREA	HEATED AREA	BED ROOMS	BATHS	PRIMARY EXTERIOR	SECONDARY EXTERIOR	HEATING	COOLING	ACTUAL YEAR BUILT	EFFECTIVE YEAR BUILT
Sketch Building 1	000100	1,635	1,403	3	2	COMMON BRICK		FORCED-AIR DUCTED	CENTRAL	1930	1985

MISCELLANEOUS DATA

DESCRIPTION	LENGTH	WIDTH	UNITS	YEAR BUILT
FP-ABV AV	0	0	1	1982
MTL UDU	10	12	120	1982

FIGURE 7.5
Section of a typical portion of an appraisal file. (Screenshot used with permission of the Schneider Corporation.)

the land use for the parcel, a description of the size and character of any improvements together with digital photographs of the improvements, other information that might assist in determining a fair value of the property, and the appraised and accessed value of the property. Often, the appraisal files are configured to serve as input into a computer-assisted mass appraisal program for automated annual adjustment of the appraised value based on comparable land sales.

It should be noted that the purchase price and date of purchase are especially important data for the appraisal file. One of the more credible methods of determining market value of property is to identify other recently sold properties with similar characteristics. Therefore, having the advantage of a database containing a record of all property sales in the jurisdiction greatly facilitates the appraisal process.

The development as well as the maintenance of both the property map and appraisal file can be a challenging effort. The process involves the identification and mapping of all property in the jurisdiction, identification of the legal owner of the each such parcel, collection of data on each parcel relevant to its value such as land use and improvements, and collection of relevant market data. For the property map, coordinate geometry and computer-assisted drafting programs are typically used for plotting of the configuration of metes and bounds descriptions and those referenced to a subdivision plat. The system must be constantly maintained by a review of all documents as they are recorded in the associated land recordation system. Those documents must be interpreted to determine the location, size, and configuration of each parcel as well as any change in ownership.

Prior to ready availability of geographic coordinate data associated with the widespread use of GPS surveying systems, separate parcel maps were typically maintained for each land section in the public land states. Since most of the parcels within each section are typically described as either an aliquot part of the section or by metes and bounds referenced to one of the corners of the section, this allowed for fairly reliable plotting of parcels within each sectional map. In nonpublic land states, however, the typical lack of a regular system of land sections made the maintenance of a property map more difficult.

Today, most municipalities require that all new subdivisions include geographic coordinate information of the subdivision plat to allow the georeferencing of all lot boundaries. Such georeferenced plats are obviously of great benefit to the maintenance of a good property map.

In addition to the land description information from recorded documents from the land recordation system for the jurisdiction, records from other sources such as the county building inspector, the clerk of court, other government agencies, and direct inspection of properties are involved in ensuring that the ownership records are current and that other factors affecting property value such as land use, improvements to land, land sale prices, and nature of the property are known. All of those factors are used in the final process of determining the appraised value of the property and ensuring that the correct land owner is taxed.

Most fiscal cadastres in the United States are based on records from an associated land recordation system. Unlike a land recordation system, which functions as a filing system, entry of data into a fiscal cadastre involves interpretation of the documents from the land recordation system to determine the current ownership of all land parcels as well as factors such as location, size, and configuration that affect the value of each parcel. In addition, unlike the typical land recordation system, a fiscal cadastre usually has a cartographic component since an up to date land parcel map is an essential component of most fiscal cadastres.

As with the land records recordation systems, many of the county fiscal cadastres have been made available for access through the Internet. As an example, with the web site for the Jefferson County, Florida, Property Appraiser's system, interested persons may search for information on a parcel using the owner's name; street address; section, township, and range; or parcel number. Information on the parcel as well as the parcel map depicting the boundaries of the parcel superimposed on an aerial photograph may then be viewed.

When using fiscal cadastral systems such as those maintained in the United States, it must be understood that there may be information revealed in a title search, boundary survey, or legal proceedings that would result in the ownership of a land parcel being other than that listed in the cadastre. In addition, it should be understood that the geographic coordinates reflected on the cadaster map may not represent those from a precise boundary survey.

7.4 The Evolving Cadastral System in the United States

7.4.1 Integration of Digital Indices and GISs

Modern technology has made significant changes in cadastral systems. One example previously discussed is the use of digital indices in land records recordation systems. That technology has provided a highly efficient approach to title searches, doing away with the need for the previously used cumbersome grantee and grantor indices.

Another example is the almost universal use of GISs for property maps associated with fiscal cadastres. This allows far more reliable maintenance of the property map and also allows systematic referencing of the file containing all of the associated valuation data with each parcel in the parcel map. Use of a GIS as a basis allows overlaying of aerial photography, topography, soil maps, wetland maps, and other data pertinent to determining the value of the property. Thus, it greatly increases the reliability of the property appraisal process.

In addition, use of a GIS as a basis allows widespread use of the parcel map for purposes other than taxation since it allows the display and analysis of multiple layers of data. Since the cadastral parcel is the basic spatial unit of most human activity, a cadastral map can serve as the ideal base map for many other aspects of society. Therefore, a parcel-based land information system may serve a wide range of users. To give a few examples: With the addition of a zoning layer, a fiscal cadastral system can serve as a base for planning and zoning. Such a system is also ideal for use as a basis for disaster impact assessment since the typical fiscal cadastre has a database of existing infrastructure and its value. It can also serve as a base map for emergency response. With 911 emergency systems to locate the caller, an emergency dispatcher or an emergency responder can, with such a system, immediately locate a residence and, with the addition of an aerial photography layer, can see the configuration of the area, alternative routes of access, locations for landing emergency helicopters, etc. With the addition of a database of locations for fire hydrants, the system can also be used to allow firefighters to see the location of the closest fire hydrants and the identity of nearby neighbors prior to arrival on the scene. In addition, such a system is a great asset for surveying, land planning, elections, school districting, agriculture, insurance, retail sales, and many other land-based activities.

As may be seen from these examples, a land information system based on a typical cadastral system has the potential of being the basis of a valuable multipurpose community resource, both for government and for the private sector. Examples of potential users in government services include planning, agricultural assistance, elections, education, environmental management, building inspection, law enforcement, taxation, health care, life assistance, utilities, transportation, and disaster management. Similarly, a multitude of potential users exist in the private sector. Thus, fiscal cadastres are rapidly becoming the basis for regional multipurpose information systems.

It should be noted that a critical requirement for such a multipurpose land information system is that the land parcel map be georeferenced. Vast amounts of data in digital format, relative to the multiple disciplines listed previously, are readily available. This includes information such as aerial photography, topography, zoning classifications, soil classification, flood zones, and wetland maps. The value of a land information system is the ability to analyze such data jointly with other data such as the location of land parcels. This can only be practically achieved by having all data being compared related to the same coordinate system. This requires that all data, including the underlying cadastral system, be georeferenced. Although most cadastral maps in the past were created using local coordinate systems, the ready availability of geographic coordinates associated with the advent of GPS technology has made georeferencing of cadastral maps more practical. The result of all this is the availability of a valuable community resource that offers significant benefits to society.

7.4.2 Improved Public Access

The concept of open records has always been an important aspect of American governance, and this is especially true of records in cadastral systems. The central concept of the U.S. land records recordation system is to serve as a means of providing public notice of claims of ownership. Therefore, the records have always been open to the public. Further, such systems have traditionally been established at the local level, which facilitated public access. Nevertheless, the traditional format of such systems and the associated indices are somewhat intimidating to a person untrained in their navigation. That complexity somewhat restricted public access. As a result, the advent of computer indexing and the Internet has had a significant effect on public access to such systems. It made searches for specific records relatively easy and it allowed access to records from any computer with Internet access.

7.4.3 Future Directions

The primary cadastral system of the United States differs from that in a number of other counties in that the juridical cadaster does not register or certify land ownership. Rather, the U.S. systems generally serve as filing systems for records of interests in land and leave assurance of the status of title and boundaries to users of that information. That approach appears to reflect the national preference and is unlikely to change in the foreseeable future. Thus the systems will probably continue to maintain the traditional roles and relationships between government and private interests. Nevertheless, as discussed in the two previous sections, many very significant changes in the nation's cadastral systems have occurred and will undoubtedly continue to occur. Many of these changes will be technology driven as with the previously discussed changes.

As an example, the use of cadastral systems as the basis for multipurpose information systems will obviously continue to expand in the future. Since most human activities are land based, the ability, provided by GIS technology, to relate the location of such activities with other information such as ownership, zoning, proximity to other important features, etc., is invaluable. Thus, more and more "layers" of information will be added to such information systems with time.

One interesting implication of the development of multipurpose parcel maps is the governmental structure housing such systems since the use of such maps for taxation is becoming only one aspect in the value and use of the maps. As more georeferenced data is added to the GIS containing the parcel maps, value to other compartments of government other than county tax assessment can actually far outweighs the original purpose. As a result, many governmental organizational changes have resulted from this trend.

The other previously mentioned change, the trend toward open access to cadastral systems via the Internet, will obviously also continue. The overall concept of open access to public records has often been the subject of controversy in the United States (Epstein and Niemann 2014). Debate has been over topics such as which, if any, records should be restricted, whether public officials should decide which records are made public, and whether or not there should be a charge for access to or copies of records. The advent of use of the Internet for accessing public records has only added to the controversy. Some of these issues have been addressed by the federal Freedom of Information Act and various state open records laws. Generally, such laws have followed the traditional openness tradition of American government.

As a result, the current trend is toward complete access to all records in both the juridical and fiscal cadastral systems in the United States. The use of digital records and links to the Internet has facilitated this trend due to reduced access and duplication costs as well as making access easier. Although some agencies continue to provide such cadastral as well as other spatial data contained in multipurpose cadastral systems, at prices beyond reproduction costs, the general trend has been to allow unlimited, free access to such data while recovering costs of developing such systems from other sources such as filing fees.

In addition to the continuation of trends toward the use of cadastral systems as a basis for multipurpose GISs and increased open access via the Internet, another change will undoubtedly be the addition of considerably more information relating to land interests in cadastral systems with time. Currently, the records in U.S. cadastral systems consist of two types of information: title records and fiscal records. Yet, there is another class of information that also has significant impact on land use and, therefore, on real estate transactions. These are administrative records associated with land use regulation and restrictions. As pointed out in earlier sections, there are many types of interests in land that fall within this category. These include interests associated with land use restrictions such as land use classification, wetland and floodplain land use restrictions, coastal zone restrictions, subdivision regulations, and many others. Information regarding such interests allows determination of what can be done with a parcel of land, including not only what can be done but also who has the power to decide. All of these interests obviously affect the use and value of a parcel and should therefore be reflected in a comprehensive land information system in addition to the land interests currently found in cadastral systems (Epstein and Niemann 2014).

With the advent of the use of GIS in cadastral systems, many of the more progressive cadastral systems in the United States have already begun to add such information to their systems. As an example, the previously mentioned Jefferson County, Florida, fiscal cadastral system has a number of classes of such information included as layers that can be selected for viewing over the property boundaries (Figure 7.6). In Figure 7.6, note the available layers that may be selected to the right side of the map itself. As may be seen, a number of them include administrative records affecting land use and value, including lakes and rivers, which may include publicly owner submerged lands; boundaries of water management districts, which regulate access to water; Federal Emergency Management Agency (FEMA) flood zones, which may restrict development; wetlands, which may restrict land use; and land use classifications, which define allowable land use.

It should be understood that the locations of areas affected by some types of administrative records are typically only approximated on cadastral maps. Examples are limits of regulated wetlands and boundaries of publically owned submerged lands under navigable waters (such as the mean high, ordinary high, or ordinary low water lines).

Such lines are typically only precisely located on a need basis since it would be prohibitively expensive to prelocate all such lines in a given region. Further, such lines, especially water boundaries, are ambulatory in nature, so precise location in advance of actual need would be futile.

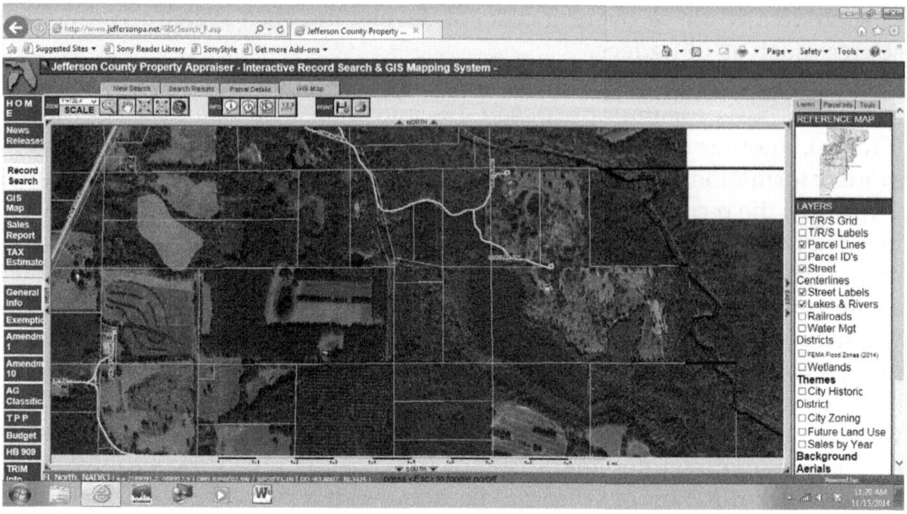

FIGURE 7.6
Jefferson County, Florida, property appraiser GIS map depicting available administrative records layers. (Screenshot used with permission of the Schneider Corporation.)

Another possible change that periodically emerges in discussions regarding the U.S. cadastral systems is a move toward a national cadastral system. As early as 1980, the National Research Council issued a report urging such a system. More recently, after the Great Recession of 2008, there were claims that a national system that incorporated parcel data into the systems that monitor mortgage and banking transactions could have prevented the crisis. Nevertheless, the United States has a long-established and well-entrenched tradition of locally maintained cadastral systems. Yet, there is no doubt that there will be trends toward the development of national standards for these systems as well as toward their linkage.

7.4.4 Impact on the Role of Land Surveying

The widespread use of cadastral systems as the basis for multipurpose GISs has had a significant effect on the role of land surveying. Land surveying practice has always been at the very heart of land tenure and cadastral systems. With functions such as the subdivision of land, preparation of plats of surveys and land descriptions, and interpretation of land descriptions for entry into cadastral maps, surveying practice has provided the basic ingredients for modern land tenure practices and cadastral systems. With the advent of georeferenced GIS, the role for surveying has expanded to include the provision of geographical coordinates in addition to the traditional products of boundary surveying. This requires knowledge of modern geodetic science as well as that of traditional cadastral surveying.

Modern GISs are capable of depicting the relative positions of features with great precision. Yet, optimum precision is rarely realized due to deficiencies in the input data. For precision when depicting the relative positions of features, all geographic coordinates in the system must be based on the same spheroid and datum. Further, modern geodetic science has recognized that the land masses of the earth are in a state of continual flux due to tectonic motion, which adds additional complexity. As a result, for a precise GIS, all coordinates entered must be based on a specific epoch as well as the same spheroid and datum. Standards must therefore be developed for data as well as for informing users of the level of precision associated with various layers of information. This means that surveyors must ensure that their work products meet those standards. These steps are critical to allow users to know the actual level of precision associated with data in such systems to prevent incorrect business and administrative decisions being made by persons unaware of the limitations.

This will not be an easy task since geographical coordinates for parcels and other types of information depicted in multipurpose systems are obtained from many different sources. These include recorded subdivision plats that have been georeferenced by a wide variety of methods with varying levels of precision, databases of geographical coordinates for U.S. Public Land Survey section corners, wetland maps and flood-prone areas generated by federal agencies, etc. Widely differing levels of precision are associated with such data.

One concern of the surveying community on the impact of trends in cadastral systems is the growing reliance on coordinates from such systems as representing the true position of property boundaries. The precise location of most boundaries may not be determined by coordinates alone. As discussed in previous chapters, physical monuments and other physical evidence has generally been held to prevail over dimensions found in land descriptions. Therefore, the location of actual boundaries may differ significantly from those reflected in descriptions and coordinates of record.

With the understanding that the precise location of land boundaries may not be reflected by cadastral maps, the demands of modern systems will no doubt result in changes in surveying practice to facilitate the best possible depiction of parcel location in cadastral systems. Such demands have already had the effect of requirements in some jurisdictions for the filing of subdivision plats in georeferenced digital formats as well as requirements for the filing of geographic coordinates for U.S. Public Land Survey corners used in surveys. Such requirements will no doubt be refined to require a formal identification of the datum and epoch of coordinates as well as for the provision of geographic coordinate information for all boundary survey products. With the modernization of cadastral systems and their linkage with all aspects of society, surveying processes will have to change to support that modernization. Surveying is inextricable linked to cadastral systems and land tenure. Therefore, changes in one will ultimately result in changes in the others.

Recommended Additional Reading on Cadastral Systems

Dale, Peter F. and John D. McLaughlin (1999). *Land Administration.* Oxford: Oxford University Press.

Epstein, Earl F. and Bernard J. Niemann Jr. (2014). *Modernizing American Land Records.* Redlands: ESRI Press.

Wilson, Donald A. (2015). *Interpreting Land Records.* Second Edition. Hoboken: John Wiley & Sons, Inc.

References

Abrams, Charles (1979). *Revolution in Land*. New York: Arno Press.

Alexander, William DeWitt (1882). *A Brief History of Land Titles in the Hawaiian Kingdom*. Honolulu: P.C. Advertiser.

Anderson, James M. and Edward M. Mikhail (1998). *Surveying Theory and Practice*. 7th Edition. New York: McGraw-Hill.

Bade, Edward S. (1940). "Titles, Points, and Lines in Lakes and Streams." *Minnesota Law Review*, Vol. 24, 305.

Blackstone – Sir William (1783). *Commentaries on the Laws of England*. London: Strahan.

Blake, Landon (2014). "Creating Superior Land Description Plats." *Professional Surveyor Magazine*.

Bureau of Land Management (2009). *Manual of Surveying Instruction*. Washington, DC: U.S. Government Printing Office.

Brown, Curtis M., Walter G. Robillard, and Donald A. Wilson (1981). *Evidence and Procedures for Boundary Location*. New York: John Wiley & Sons.

Cadle, Farris W. (1991). *Georgia Land Surveying History and Law*. Athens: The University of Georgia Press.

Chandler, Alfred N. (1945). *Land Title Origins: A Tale of Force and Fraud*. New York: Robert Schalkenbach Foundation.

Clawson, Marion (1964). *Man and Land in the United States*. Lincoln: University of Nebraska Press.

Clawson, Marion (1968). *The Land System of the United States*. Lincoln: University of Nebraska Press.

Coastal States Organization (1997). *Putting the Public Trust Doctrine to Work*. 2nd Edition.

Cole, George M. (1990). "*Significance of Early State Land Records to Modern Land Use*" (unpublished directed individual study). Tallahassee: Florida State University, Department of Geography.

Cole, George M. (1991). "Tidal Water Boundaries." *Stetson Law Review*, Vol. XX, Nos. 1 & 2, Fall 1990 & Spring 1991, 165–176.

Cole, George M. (1997). *Water Boundaries*. New York: John Wiley & Sons.

Cole, George M. (2007). *Hydrology-Based Wetland Delineation* (doctorial dissertation). Tallahassee: Florida State University.

Cole, George M. (2011). "Sea Level Measurements and their Applications." *Proceedings*, SaGES Conference, Mayaguez, Puerto Rico.

Dale, Peter F. and John D. McLaughlin (1999). *Land Administration*. Oxford: Oxford University Press.

De Soto, Hernando (2000). *The Mystery of Capital*. New York: Basic Books (Perseus Book Group).

Dekker, Henri A.L. (2003). *The Invisible Line—Land Reform, Land Tenure, Security, and Land Registration*. Aldershot, England: Ashgate Publishing, Ltd.

Epstein, Earl F. and Bernard J. Niemann Jr. (2014). *Modernizing American Land Records*. Redlands: ESRI Press.

Gipson, Lawrence Henry (1915). "Criminal Codes of Pennsylvania." *Journal of Criminal Law and Criminology*, Vol. 6, Issue 3, Article 2, 323.

Glick, Thomas F. (2005). *Islamic and Christian Spain in the Middle Ages*. 2nd Edition. Boston: Brill Publishing.

Glick, Thomas F., Steven J. Livesey, and Faith Willis (2005). *Medieval Science, Technology and Medicine*. New York: Taylor & Francis.

Griffin, Robert J. (1960). "Replacement and Apportionment as Surveying Methods for Re-establishing Property Corners." *Marquette Law Review*, Vol. 43, 484–510.

Hale, M. (1666). *De Jure Maris*. Reprinted in S. Moore (1888). *A History of the Foreshore and the Law Relating Thereto*, 3rd Edition London: Stevens & Haynes.

Hicks, Stacy D., Henry A. Debaugh, and Leonard E. Hickman Jr. (1983). *Sea Level Variations for the United States 1855–1980*. Silver Springs, MD: U.S. Department of Commerce, NOAA, NOS.

Hilliard, Sam B. (1973). *An Introduction to Land Survey Systems in the Southeast*. West Georgia College, Studies in the Social Sciences, June 1973, 1–15.

Hodler, Thomas W. (1986). *The Atlas of Georgia*. Athens: The University of Georgia Institute of Community and Area Development.

Hogue, Arthur R. (1966). *Origins of the Common Law*. Bloomington: Indiana University Press.

Johnson, J.B. and Leonard S. Smith (1913). *The Theory and Practice of Surveying*. 17th Edition. New York: Wiley & Sons.

Larsson, Gerhard (1991). *Land Registration and Cadastral Systems*. Harlow, Essex, England: Longman Scientific and Technical.

Linklater, Andro (2002). *Measuring America*. New York: Penguin Group.

Locke, John (1689/1952). *Second Treatise of Government*. New York: The Liberal Arts Press.

Maitland, Fredrick William and Francis Charles Montague (1915). *A Sketch of Legal English History*. New York: G.P. Putnam's Sons.

Maloney, Frank E. and Richard C. Ausness (1974). "The Use and Legal Significance of the Mean High Water Line in Coastal Boundary Mapping." *The North Carolina Law Review*, 185–273.

McEntyre, John G. (1978). *Land Survey Systems*. New York: John Wiley & Sons.

Morisset, Mason D. (2001). *Tribal Rights and their Effect on our Concept of Property Rights in the Northwest*. Available at: www.msaj.com/papers.

Mulford, A.C. (1912). *Boundaries and Land Landmarks*. Boston: F.H. Gibson Company.

Platt, Rutherford H. (2014). *Land and Society*. Washington, DC: Island Press.

Powelson, John P. (1988). *The Story of Land*. Cambridge: The Lincoln Institute of Land Policy.

Robillard, Walter G. and Donald A. Wilson (2011). *Brown's Evidence and Procedures for Boundary Location*. 6th Edition. New York: John Wiley & Sons.

Robillard, Walter G. and Donald A. Wilson (2014). *Brown's Boundary Control and Legal Principles*. 7th Edition. New York: John Wiley & Sons.

Sandars, Thomas C. (1874). *The Institutes of Justinian*. London: Longmans, Green & Co.

Schachner, Nathan (1951). *Thomas Jefferson—A Biography*. New York: Thomas Yoseloff.

Scott, Samuel P. (1931). *Las Siete Partidas*. New York: Commerce Clearing House.

Shick, Blair C. and Irving H. Plotkin (1978). *Torrens in the United States*. Lexington: D.C. Heath & Company.

Shofner, Jerrell H. (1976). *History of Jefferson County*. Tallahassee: Sentry Press.

Truesdell, W.A. (1908). *The Rectangular System of Surveying*. New York: Journal of the Association of Engineering Science.

Woodward, William E. (1926). *George Washington—The Image and the Man*. New York: Boni and Liveright.

Index

Page numbers with f and n refer to figures and footnotes, respectively.

D

G

Y

An environmentally friendly book printed and bound in England by www.printondemand-worldwide.com

PEFC Certified

This product is from sustainably managed forests and controlled sources

www.pefc.org

PEFC/16-33-415

This book is made of chain-of-custody materials; FSC materials for the cover and PEFC materials for the text pages.